SpringerBriefs in Applied Sciences and Technology

Reliability

Series editors

Cher Ming Tan, Singapore, Singapore
Xuejun Fan, Beaumont, TX, USA

More information about this series at http://www.springer.com/series/11543

Jiann-Shiun Yuan

CMOS RF Circuit Design for Reliability and Variability

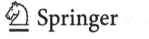 Springer

Jiann-Shiun Yuan
Department of Electrical and Computer
  Engineering
University of Central Florida
Orlando, FL
USA

ISSN 2191-530X   ISSN 2191-5318   (electronic)
SpringerBriefs in Applied Sciences and Technology
ISSN 2196-1123   ISSN 2196-1131   (electronic)
SpringerBriefs in Reliability
ISBN 978-981-10-0882-5   ISBN 978-981-10-0884-9   (eBook)
DOI 10.1007/978-981-10-0884-9

Library of Congress Control Number: 2016938045

Printed on acid-free paper

This Springer imprint is published by Springer Nature
The registered company is Springer Science+Business Media Singapore Pte Ltd.

Contents

Chapter 1
Introduction

It is well known that wireless transceivers are implemented in mobile devices such as smart phones, laptops, tablets, etc. Wireless transceivers are also critical circuit blocks for sensors in the Internet of Things (IoT) era. IoT is being represented as a worldwide network interconnecting things/objects. IoT is a kind of technology that realizes the communication and information exchange between machine and human and machine by embedded RFID, GPS, and sensors technologies into physical equipment, and achieve transition, cooperation, processing of information according to some protocols and so that achieve the goal of intelligent identification, tracking, monitoring, computing, and management. IoT is made up of sensing layer, network layer, and application layer. Sensing layer is responsible for accumulation of data and information. Network layer realizes the management of connection of network and data and transmits information to application layer. Application layer processes information in order to realize monitoring, identification, control, and other functions. Network layer mainly guarantees the connection of network. It can support the network protocols of internet and provide efficient channel for voice and data. To sum up, IoT is a combination of many kinds of networking technologies, and at the same time, IoT cannot be developed without the support of communication network.

Clearly, wireless technologies are very important in IoT area due to the convenient and low cost wireless connections between IoT nodes. RF transceiver is the critical block in wireless nodes and consumes the majority of energy. A typical super-heterodyne architecture transceiver is widely used in RF transceivers with better sensitivity and higher gain. For a super-heterodyne topology in RF transceiver, for example in the receiver (RX) path, the RF signal coming from the antenna and RF switch goes to the front-end low-noise amplifier (LNA). The RF signal is amplified by the LNA and down converted to the intermediate frequency (IF) signal using the mixer and local oscillator (LO). The IF signal then passes through the analog-to-digital (A/D) converter for base band digital signal processing. On the other hand, for the transmitter (TX) path, the digital signal passed through the digital-to-analog (D/A) converter to produce the analog signal. The IF

© The Author(s) 2016
J.-S. Yuan, *CMOS RF Circuit Design for Reliability and Variability*,
SpringerBriefs in Reliability, DOI 10.1007/978-981-10-0884-9_1

is then up-converted to RF signal using the up-converting mixer and LO at desired frequency. The RF signal is amplified by the power amplifier (PA). The RF switch connects the large-signal RF waveform to the antenna for signal transmission.

RF transceiver circuits including low noise amplifiers, mixers, oscillators, and power amplifiers are usually made using mixed-signal CMOS technology. CMOS is an ideal candidate for high density, low cost, low power, and high integration chip solution. Today, silicon CMOS are scaled down to 22 nm and beyond to increase density and performance further. The well-known reliability mechanisms such as hot carrier injection (HCI), negative bias temperature instability (NBTI), and gate oxide breakdown (GOB) become very important knowledge for the design of advanced RF and digital circuits. For state-of-the-art nanoscale circuits and systems, the local device variation and uncertainty of signal propagation time have become crucial in the determination of system performance and reliability. Yield analysis and optimization, which take into account the manufacturing tolerances, model uncertainties, variations in the process parameters, and aging factors are known as indispensable components of the circuit design procedure. Therefore, circuit designers, device engineers, and graduate students need to have clear understanding on how device reliability issues affect the RF circuit performance subjected to operation aging and process variations. This book is unique to explore typical reliability issues in the device and technology level and then to examine their impact on RF wireless transceiver circuit performance. Analytical equations, experimental data, mixed-mode device, and circuit simulation results will be given for clear illustration.

Chapter 2
CMOS Transistor Reliability and Variability Mechanisms

Due to aggressive scaling in device dimensions for improving speed and functionality, CMOS transistors in the nanometer regime have resulted in major reliability issues due to high electric field phenomenon. These include hot carrier injection (HCI) [1, 2], gate oxide breakdown (BD) [3, 4], and negative bias temperature instability (NBTI) [5, 6]. These reliability mechanisms cause the MOS transistor parameter drifts; namely, threshold voltage shift and mobility degradation. A brief discussion on the MOS device reliability is described as follows.

2.1 Hot Electron Effect

When the electric field at the drain edge of the MOS transistor is very high, avalanche breakdown may occur. Impact ionization in the drain depletion region generates many energized electrons. These high energy carriers may damage interfacial layer and create interface traps and oxide trapped charges [7] which degrade device parameters such as an increase in threshold voltage. Figure 2.1 displays the drain current degradation versus drain-source voltage subjected to different stress times. At given drain-source voltage V_{DS} and gate-source voltage V_{GS}, the drain current decreases with stress time as shown in Fig. 2.1.

2.2 Gate Oxide Breakdown

High electric field across the gate insulator could induce time-dependent dielectric breakdown. The formation of random defects and conduction path within the gate dielectric increases the gate leakage and noise. For ultrathin gate oxide transistors under constant gate voltage stress, the soft breakdown could be observed before hard breakdown [8]. Compared with hard breakdown (HBD), SBD becomes more

© The Author(s) 2016
J.-S. Yuan, *CMOS RF Circuit Design for Reliability and Variability*,
SpringerBriefs in Reliability, DOI 10.1007/978-981-10-0884-9_2

Fig. 2.1 Drain current
degradation due to hot
electron stress (© IEEE)

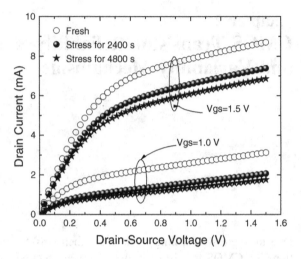

Fig. 2.1 Drain current degradation due to hot electron stress (© IEEE)

Fig. 2.2 Normalized I_g
versus stress time. The nMOS
is under positive or negative
constant gate bias (© IEEE)

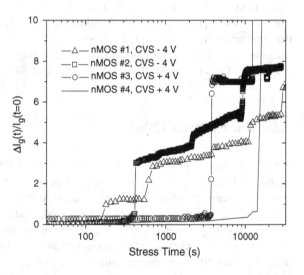

Fig. 2.2 Normalized I_g versus stress time. The nMOS is under positive or negative constant gate bias (© IEEE)

prevalent for thinner oxides and for oxide stress at relatively lower voltages. In addition, hot carrier injection could trigger more SBD in addition to conventional Fowler–Nordheim (FN) tunneling [9].

Figure 2.2 shows the normalized gate leakage current as a function of stress time under constant voltage (CVS). The gate soft breakdown degrades the threshold voltage and mobility of the MOSFET as observed by the current–voltage characteristics [10].

2.3   Negative Bias Temperature Instability

Negative bias temperature instability is related to a build up of positive charges occurring at the Si/SiO_2 interface or in the oxide layer for p-channel transistors under negative gate bias. The reaction–diffusion model [11] illustrates the holes in the inversion layer of pMOSFETs reacting with the Si–H bonds at the SiO_2/Si interface. The hydrogen species diffuse away from the interface toward the polysilicon gate. This causes the threshold voltage instability of pMOSFETs. The NBTI effect is enhanced at higher temperatures. Note that NBTI is a degradation of transistor performance for pMOSFETs, where positive bias temperature instability (PBTI) transistor occurs for nMOSFETs with high-k dielectrics [12].

To investigate the oxide breakdown and hot electron effect on the nMOS transistors at various stress conditions, accelerated DC voltage stress is employed. Figure 2.3 shows the drain current versus drain-source voltage and Fig. 2.4 displays the transconductance versus gate voltage of the 65 nm nMOS during 220 min of hot electron stress at $V_{GS} = 0.35$ V and $V_{DS} = 2.0$ V. At high drain-source voltage, hot carrier injection occurs because of high electric field and impact ionization at the drain region of MOSFETs. Again, these high energy carriers may introduce damage by creating interface traps and oxide trapped charges and can cause degradation of device parameters such as an increase in threshold voltage and a decrease in transconductance. At a given drain voltage, the drain current decreases with stress time and at a given gate voltage, the transconductance decreases with stress time due to hot electron degradation.

The 65 nm NMOS is also measured under gate oxide stress at $V_{GS} = 2.9$ V and $V_{DS} = 0$ V. The results are shown in Fig. 2.5. After significant oxide stress effect resulting from high gate voltage, the transconductance shifts down rapidly in the initial 60 min as seen in Fig. 2.5. The off-state stress effect is evaluated in Fig. 2.6.

Fig. 2.3 I_D-V_D at different stress times (DC stress at $V_{GS} = 0.35$ V ad $V_{DS} = 2.0$ V)

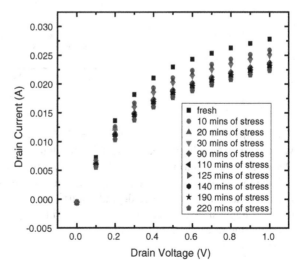

Fig. 2.4 Transconductance at different stress times (DC stress at $V_{GS} = 0.35$ V ad $V_{DS} = 2.0$ V)

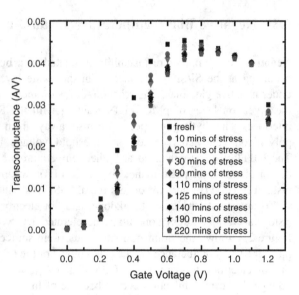

Fig. 2.5 Transconductance versus gate voltage (oxide stress at $V_{GS} = 2.9$ V ad $V_{DS} = 0$ V)

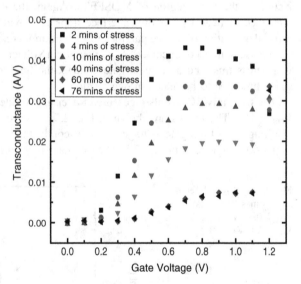

The 65 nm nMOS was stressed at $V_{GS} = 0$ V ad $V_{DS} = 2.8$ V. High drain-source voltage results in high electric field in the drain region, which may trigger hot electron injection into the gate oxide to degrade the drain current. High drain-gate voltage may also induce gate oxide breakdown close to the drain edge. As shown in Fig. 2.6, the transconductance degrades quickly after only 30 min of off-state high drain voltage stress. After 30 min of stressing, the transconductance collapses possibly due to oxide hard breakdown accelerated by hot electron injection during off-state.

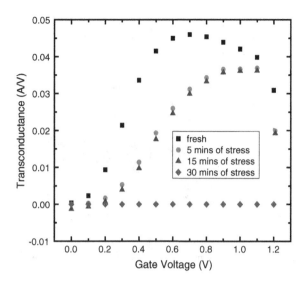

Fig. 2.6 Transconductance at different stress times (DC stress at $V_{GS} = 0$ V ad $V_{DS} = 2.8$ V)

2.4   Process Variability

Process variations were originally considered in die-to-die variations. For nanoscale transistors, intra die variations are posing the major design challenge as technology node scales. The intrinsic device parameter fluctuations that result from process uncertainties have substantially affected the device characteristics. Process variability comes from random dopant fluctuation, line edge roughness, and poly gate granularity [13, 14]. The threshold voltage fluctuation due to random doping profile is approximated as [15]:

$$\sigma^2_{Vt,\text{doping}} = \frac{2q^2 t^2_{\text{ox}}}{WL\varepsilon^2_{\text{ox}}} \int_0^{W_D} N_A(x)(1 - \frac{x}{W_D})^2 dx \qquad (2.1)$$

where q is electron charge, t_{ox} is the oxide capacitance, W is the channel width, L is the channel length, ε_{ox} is the oxide permittivity, and N_A is the acceptor doping. With shrinking of gate length, the deviation of threshold voltage is expected to be larger.

A computational effective device simulator [16] is used into demonstrate random doping fluctuation effect on the MOSFET model parameter variation. A 22 nm LDD NMOS transistor is constructed as an example to illustrate the threshold voltage fluctuation. From Fig. 2.7, it is seen that the acceptor dopant causes positive V_T fluctuation with peak value of 0.0045 V located around the center of the channel. Due to the random doping fluctuation, the standard deviation (STD) of V_T for the 22 nm MOSFET is computed to be 0.031 V or its corresponding spread (STD/Mean) of 6.9 %.

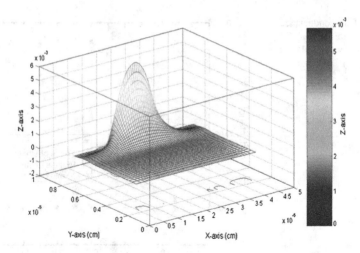

Fig. 2.7 Sensitivity function distribution of V_T versus acceptor (© IEEE)

CMOS technology continues device scaling for high integration. However, as feature sizes shrink and chip designers attempt to reduce supply voltage to meet power targets in large multi-core systems, parameter variations are becoming a serious problem. Parameter variations can be broadly classified into device variations incurred due to imperfections in the manufacturing process and environmental variations and on-die temperature and supply voltage fluctuations. Smaller feature size further makes CMOS circuits more vulnerable to process, supply voltage, and temperature (PVT) variability. Large design margin is then needed to insure circuit robustness against reliability issues. Using PVT and long-term reliability resilience design is becoming an essential design requirement for the future technology nodes and may reduce overdesign, while increasing yield and circuit robustness.

References

1. Park J-T, Lee B-J, Kim D-W, Yu C-G, Yu H-K (2000) RF performance degradation in nMOS transistors due to hot carrier effects. IEEE Trans Electron Devices 47(5):1068–1072
2. Pantisano L, Schreurs D, Kaczer B, Jeamsaksiri W, Venegas R, Degraeve R, Cheung KP, Groeseneken G (2003) RF performance vulnerability to hot carrier stress and consequent breakdown in low power 90 nm RFCMOS. In: IEDM Technical Digest, pp 181–184
3. Depas M, Nigam T, Heyns MM (1996) Soft breakdown of ultra-thin gate oxide layers. IEEE Trans Electron Device 1499–1504
4. Yu C, Yuan JS (2007) CMOS device and circuit degradations subject to HfO$_2$ gate breakdown and transient charge-trapping effect. IEEE Trans Electron Devices 59–67
5. Stathis JH, Zafar S (2006)The negative bias temperature instability in MOS devices: a review. In: Microelectronics Reliability, pp 270–286
6. Jeppson KO, Svensson CM (1977) Negative bias stress of MOS devices at high electric fields and degradation of NMOS device. J Appl Phys 2004–2016

7. Ang DS, Ling CH (1999) The role of electron traps on the poststress interface trap generation in hot-carrier stressed p-MOSFETs. IEEE Trans Electron Devices 46(4):738–746

8. Alam MA, Weir B, Bude J, Silverman P, Monroe D (1999) Explanation of soft and hard breakdown and its consequences for area scaling. Int Electron Devices Meet 449–452

9. Huang J, Chen TP, Tse MS (2001) Study of edge charge trapping in gate oxide caused by FN and hot-carrier injection. Conf Optoelectron Microelectron Mater Devices 409–412

10. Liu Y, Sadat A, Yu C, Yuan JS (2001) RF performance degradation in pMOS transistors due to hot carrier and soft breakdown effects. In: Topical Meeting on Silicon Monolithic Integrated Circuits in RF Systems, pp 309–310

11. Alam MA, Kufluoglu H (2006) Theory of interface-trap-induced NBTI degradation for reduced cross section MOSFETs. IEEE Trans Electron Devices 1120–1130

12. Lee KT, Kang CY, Yoo OS, Choi R, Lee BH, Lee JC, Lee HD, Jeong YH (2008) PBTI-associated high-temperature hot carrier degradation of nMOSFETs with metal-gate/high- k dielectrics. IEEE Electron Device Lett 389–391

13. Li Y, Huang CH, Li TY (2009) Random-dopant-induced variability in nano-CMOS devices and digital circuits. IEEE Trans Electron Devices 1588–1597

14. Kokkoris G, Constantoudis V, Gogolides E (2009) Nanoscale roughness effects at the interface of lithogrphy and plasma etching: modeling of line-edge-roughness transfer during plasma etching. IEEE Trans Plasma Sci 1705–1714

15. Stolk PA, Widdershoven FP, Klaassen DBM (1998) Modeling statistical dopant fluctuations in MOS transistors. IEEE Trans Electron Devices 1960–1971

16. RandFlux, RandFlux v.6, User's manual, Florida State University. P. User Manual v. 0.6

Chapter 3
Low-Noise Amplifier Reliability

Low-noise amplifier (LNA) is used in the first stage of wireless transceiver to amplify signal with low noise. The gain and noise performances of the LNA have a strong effect on the noise performance of the whole system [1]. Since semiconductor circuits are sensitive to device parameter variations, a robust design in the wireless transceiver to maintain the required performance over a wide range of stress conditions is desirable.

3.1 Analytical Equations

The low-noise amplifier design can be briefly described as follows. The input matching is important to determine the noise figure of the circuit. The input gate inductor, source inductor, input transistor's transconductance, and gate capacitance are designed to provide the input matching to 50 Ω at the operating frequency. The LNA can be in cascode and/or cascade topologies. The cascade (multistage) topology increases the overall small-signal power gain. The cascode topology improves the input and output isolation. In the cascade topology, the first stage transistor usually determines the overall noise figure, while the last stage transistor dominates the linearity (IP3) of the LNA [2]. The minimum noise figure of a single stage LNA can be written as

$$\text{NF}_{\min} = 1 + 2\pi K_f (C_{\text{GS}} + C_{\text{GD}}) \sqrt{\frac{R_S + R_D}{g_m}} \tag{3.1}$$

where K_f is the Fukui constant, C_{GS} and C_{GD} are the gate-source capacitance and gate-drain capacitance, g_m is the transconductance of the input transistor, and R_S

© The Author(s) 2016
J.-S. Yuan, *CMOS RF Circuit Design for Reliability and Variability*,
SpringerBriefs in Reliability, DOI 10.1007/978-981-10-0884-9_3

and R_D are the source resistance and drain resistance, respectively. Note that NF_{min} is the minimum noise figure a LNA can achieve without taking into account noise from the biasing circuitry and matching components such as input and output inductors and capacitors.

Using (3.1) the minimum noise figure sensitivity is derived as

$$\Delta NF_{min} = \frac{\partial NF_{min}}{\partial g_m} \Delta g_m = -\pi K_f (C_{GS} + C_{GD}) \sqrt{\frac{R_S + R_D}{g_m^3}} \Delta g_m. \qquad (3.2)$$

In (3.2) the transconductance change Δg_m can be related to threshold voltage shift ΔV_T and electron mobility degradation $\Delta \mu_n$ as follows:

$$\Delta g_m = \frac{\partial g_m}{\partial V_T} \Delta V_T + \frac{\partial g_m}{\partial \mu_n} \Delta \mu_n. \qquad (3.3)$$

The drain current of the input transistor in saturation can be expressed as

$$I_{DS} = \frac{\mu_n C_{OX} W}{2L} \frac{(V_{GS} - V_T)^2}{1 + \theta(V_{GS} - V_T)} (1 + \lambda V_{DS}), \qquad (3.4)$$

where μ_n is the electron mobility, V_T is the threshold voltage, C_{OX} is the oxide capacitance per unit area, W is the channel width, L is the channel length, λ is the channel length modulation factor, and θ accounts for the mobility degradation due to vertical electrical field from the gate. The transconductance of MOSFET can be derived as

$$g_m = \frac{\partial I_{DS}}{\partial V_{GS}} = \frac{\mu_n C_{OX} W}{2L} \left\{ \frac{2(V_{GS} - V_T)}{1 + \theta(V_{GS} - V_T)} - \frac{\theta(V_{GS} - V_T)^2}{[1 + \theta(V_{GS} - V_T)]^2} \right\} (1 + \lambda V_{DS}).$$
$$(3.5)$$

Key transistor parametric sensitivity on the transconductance is given by

$$\frac{\partial g_m}{\partial V_T} = \frac{\mu_n C_{OX} W}{2L} \left\{ \frac{-2}{[1 + \theta(V_{GS} - V_T)]^3} \right\} (1 + \lambda V_{DS}), \qquad (3.6)$$

$$\frac{\partial g_m}{\partial \mu_n} = \frac{C_{OX} W}{2L} \left\{ \frac{2(V_{GS} - V_T)}{1 + \theta(V_{GS} - V_T)} - \frac{\theta(V_{GS} - V_T)^2}{[1 + \theta(V_{GS} - V_T)]^2} \right\} (1 + \lambda V_{DS}). \qquad (3.7)$$

Inserting (3.6) and (3.7) into (3.3) gives

$$
\Delta g_m = -\frac{\mu_n C_{OX} W}{L} \left\{ \frac{1}{[1+\theta(V_{GS}-V_T)]^3} \right\} (1+\lambda V_{DS}) \Delta V_T
$$
$$
+ \frac{C_{OX} W}{2L} \left\{ \frac{2(V_{GS}-V_T)}{1+\theta(V_{GS}-V_T)} - \frac{\theta(V_{GS}-V_T)^2}{[1+\theta(V_{GS}-V_T)]^2} \right\} (1+\lambda V_{DS}) \Delta \mu_n.
$$

$$(3.8)$$

Using (3.1) and (3.2) the normalized minimum noise figure can be written as

$$
\frac{\Delta NF_{min}}{NF_{min}-1} = -\frac{1}{2} \frac{\Delta g_m}{g_m}. \tag{3.9}
$$

Inserting (3.5) and (3.8) into (3.9) results in

$$
\frac{\Delta NF_{min}}{NF_{min}-1} = \frac{V_T}{2(V_{GS}-V_T)+3\theta(V_{GS}-V_T)^2+\theta^2(V_{GS}-V_T)^3} \frac{\Delta V_T}{V_T} - \frac{1}{2}\frac{\Delta \mu_n}{\mu_n}.
$$

$$(3.10)$$

Equation (3.10) indicates that when the threshold voltage increases ($\Delta V_T > 0$) and/or electron mobility decreases ($\Delta \mu_n < 0$), the normalized minimum noise figure will increase. Note that in (3.10) ΔV_T and $\Delta \mu_n$ could be from transistor aging and/or process variations. The analytical equations developed provide physical insight into how the stress effects such as an increase of threshold voltage and a decrease of electron mobility affect the LNA noise performance.

The effects of hot carrier stress on the characteristics of a fully integrated LNA designed at a few giga hertz frequency operation using 0.18 μm CMOS technology were investigated experimentally [3]. After electrical stress, the threshold voltage of the misfit increased and the channel electrons' mobility decreased leading to a reduction of the biasing currents. Small-signal parameter such as transconductance g_m is decreased and output conductance g_{ds} is increased. There was no change measured in the gate-source and gate-drain capacitances. The LNA's power gain S_{21} decreased after stress mainly due to the drop in the transconductance and, to some extent, due to the increase in their output conductance. The input and output matching of the LNA are slightly deteriorated with stress due to the change in transconductance and output conductance of the transistors.

3.2 LNA Fabrication and Experimental Data

To evaluate the LNA stress effects at the millimeter-wave frequency range (60–70 GHz), a three-stage low-noise amplifier shown in Fig. 3.1 was fabricated using 65 nm CMOS technology [4]. The LNA is operating at millimeter-meter frequency which could be used for high speed, low cost home entertainment, imaging, and automotive radar systems applications. In Fig. 3.1 transistors $M1$, $M2$, and $M3$ are for the first, second, and third stage, respectively. $M4$ is the cascode transistor in the third stage to reduce the output transistor voltage stress. The first stage transistor usually determines the overall noise figure, while all three stages contribute to the small-signal power gain. The source inductor of the LNA is used to reduce noise figure, while maintaining the small-signal power gain. The input gate inductor, source inductor, and $M1$'s transconductance and gate capacitance simultaneously provide the input matching to 50 Ω. The coupling capacitors C_1, C_2, C_3, and C_4 allow the millimeter-wave signal to go through the coupling capacitors, while blocking the DC signal to the MMW input and output.

The optimization of the different stages was performed using the methodology in [5]. The input, interstage, and output matching networks were implemented using metal-oxide-metal capacitors and multilayer coplanar waveguide transmission lines. The values of $L1$ and $C1$ were calculated for optimum impedance and noise figure matching. The fabricated die is shown in Fig. 3.2. More fabrication detail of this

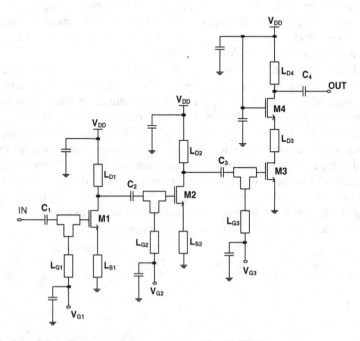

Fig. 3.1 Schematics of a three-stage low-noise amplifier (© IEEE)

Fig. 3.2 Die photo of the millimeter-wave low-noise amplifier (© IEEE)

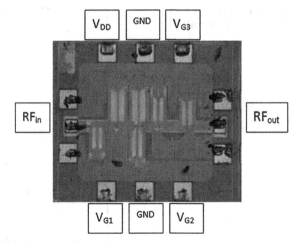

LNA can be found in [5]. In this diagram the millimeter-wave input (GSG) pad is on the left, the output signal (GSG) pad is on the right, the top DC pad is used for V_{DD}, GND, V_{G3}, and the bottom DC pad is for V_{G1}, GND, and V_{G2}.

Figure 3.3 demonstrates the noise figure versus frequency from 60 to 65 GHz [6]. The LNA is biased at V_{G1} = 0.75 V, V_{G2} = 0.75 V, V_{G3} = 0.75 V, and V_{DD} = 1.5 V. The V_{DD} is increased to 2.2 V during accelerated voltage stress. As seen in Fig. 3.4 the LNA minimum noise figure occurs at 63.5 GHz due to good transconductance and input matching at that frequency point. The noise figure increases with stress time due to enhanced hot carrier effect. At high drain voltage, carriers in the channel gain high energy in the pinch-off region and cause an avalanche effect. The collisions of hot carriers at the silicon and SiO_2 interface leads

Fig. 3.3 Noise figure versus frequency (© IEEE)

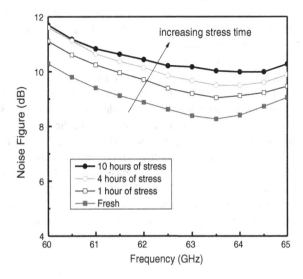

Fig. 3.4 Current from V_{DD}
versus stress time (© IEEE)

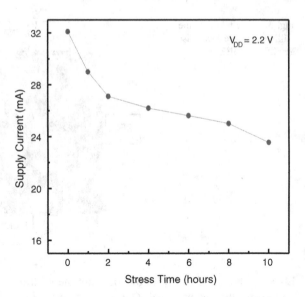

to the generation of dangling bonds, also known as interface traps. This decreases
the drain current of the nMOS, as evidenced by the measured supply current
reduction over time in Fig. 3.4. As in Fig. 3.4 the supply current decreases rapidly
in the beginning of the first two to three hours and then gradually saturates after-
ward. Hot carrier effect also decreases the transconductance of the transistor. The
reduction in transconductance results in an increase of the noise figure. After 10 h
of V_{DD} overstress, the minimum noise figure increases about 2 dB. Note that the
more the stress effect, the flatter the noise figure curve is over the frequency range
from 60 to 65 GHz.

To further investigate the hot carrier effect on the LNA, 65 nm transistor under
accelerated voltage stressing is examined. The experimental data demonstrate that
the drain current degrades significantly at $V_{GS} - V_T = 0.35$ V and $V_{DS} = 2.0$ V
(close to the maximum substrate current mode of HCI stress). Previous studies [7,
8] indicate that the maximum channel hot electron effect of drain current degradation occurs at $V_{GS} = V_{DS}$. According to [9], however, much server degradation of
high frequency noise appears under the maximum substrate current stress mode
attributing to the much larger population of shallow Si/SiO_2 interface traps, which
significantly contribute to channel noise in the high frequency range. Note that the
drain current is a DC parameter, while NF_{min} is a RF parameter.

The input reflection coefficient S_{11} versus frequency is displayed in Fig. 3.5. It is
seen in Fig. 3.5 that the input matching point has been shifted significantly from
68.5 to 70.5 GHz after 10 h of voltage stress due to disruption or change of input
matching. The small-signal power gain S_{21}, also shifts downward to the right, as
shown in Fig. 3.6. The small-signal power gain decreases with stress time due to

Fig. 3.5 S_{11} versus frequency
for different stress times
(© IEEE)

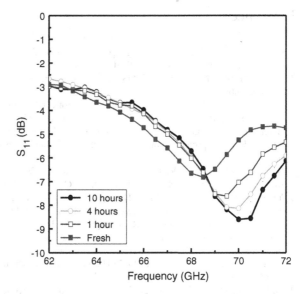

Fig. 3.6 S_{21} versus frequency
for different stress times
(© IEEE)

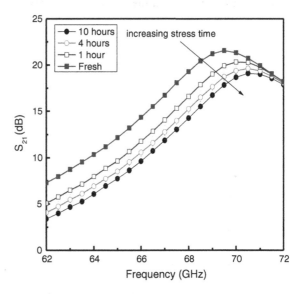

reduction of transconductance after hot electron degradation. The maximum
small-signal power gain drops more than 3 dB after 10 h of V_{DD} overstress. The
experimental data in the millimeter-wave frequency regime show that the noise
figure increases (~ 2 dB) and small-signal power gain S_{21} decreases (~ 3 dB) after
significant HCI stress due to transconductance degradation as evidenced by 65 nm
individual transistor overstress measurement.

References

1. Cabric D, Chen M, Sobel D, Yang J, Broderson RW (2005) Future wireless systems: UWB, 60 GHz and cognitive radios. In: IEEE custom integrated circuits conference, pp 793–796
2. Lee TH (1998) The design of CMOS radio-frequency integrated circuits. Cambridge University Press, Cambridge
3. Naseh S, Deen MJ, Chen C-H (2005) Effects of hot-carrier stress on the performance of CMOS low-noise amplifiers. IEEE Trans Device Mater Reliab 5:501–508
4. Berenguer R, Liu G, Xu Y (2010) A low power 77 GHz low noise amplifier with an area efficient RF-ESD protection in 65 nm CMOS. IEEE Microwave Wireless Compon Lett 20:678–680
5. Khanpour M, Tang KW, Garcia P, Voinigescu SP (2008) A wideband W-band receiver front-end in 65 nm CMOS. IEEE J Solid State Circ 43:1717–1730
6. Yuan JS, Xu Y, Yen S-D, Bi Y, Hwang G-W (2014) Hot carrier injection stress effect on a 65 nm LNA at 70 GHz. IEEE Trans Device Mater Reliab 3:931–934
7. Amat E, Kauerauf R, Degraeve R, Keersgieter An De, Rodriguez R, Nafria M, Aymerich X, Groeseneken G (2009) Channel hot-carrier degradation in short-channel transistors with high-k metal gate stacks. IEEE Trans Device Mater Reliab 425–430
8. Guerin C, Huard V, Bravaix A (2007) Hot-carrier damage from high to low voltage using the energy-driven framework. Microelectron Eng 84:1938–1942
9. Su H, Wang H, Liao H, Hu H (2012) Degradation of high-frequency noise in nMOSFETs under different modes of hot-carrier stress. IEEE Trans Electron Devices 59:3078–3083

Chapter 4
Power Amplifier Reliability

4.1 Class AB Power Amplifier

The advance in CMOS technology for high frequency applications has made it a natural choice for integrated, low-cost RF power amplifiers for wireless communications ICs. Depending on its conduction angle θ, the power amplifier is considered as the Class A, B, or C mode of operation. For example, the conduction angle of the class A PA is equal to 2π, the class B PA is π, and the class C PA is less than π. For the class AB PA the conduction angle is between π and 2π. The output power and efficiency of the PA can be written as [1]

$$P_{\text{out}} = \frac{1}{2}(V_{DD} - V_{DSAT})\frac{I_m}{2\pi}(\theta - \sin\theta) \tag{4.1}$$

$$\eta = \frac{V_m}{V_{DD}}\frac{\theta - \sin\theta}{4(\sin(0.5\theta) - 0.5\theta\cos(0.5\theta))} \tag{4.2}$$

where V_{DD} is the supply voltage, V_{DSAT} is the knee voltage, I_m is the maximum drain current in the input transistor, and V_m is the maximum output voltage. The efficiency is defined as P_{out}/P_{DC}, where P_{DC} is the DC power supplied to the circuit. η reaches the maximum when V_m is approaching V_{DD}.

Using the partial derivative approach, the change of output power and efficiency can be normalized to as follows:

$$\frac{\Delta P_{\text{out}}}{P_{\text{out}}} \approx \frac{\Delta I_m}{I_m} + \frac{\Delta\theta}{f_1(\theta)} \tag{4.3}$$

$$\frac{\Delta\eta}{\eta} \approx \frac{\Delta V_m}{V_m} + \frac{\Delta\theta}{f_2(\theta)} \tag{4.4}$$

© The Author(s) 2016
J.-S. Yuan, *CMOS RF Circuit Design for Reliability and Variability*,
SpringerBriefs in Reliability, DOI 10.1007/978-981-10-0884-9_4

where

$$f_1(\theta) = \frac{\theta - \sin\theta}{1 - \cos\theta}$$

$$f_1(\theta) = \left(\frac{1 - \cos\theta}{\theta - \sin\theta} - \frac{\theta}{4}\frac{\sin 0.5\theta}{\sin 0.5\theta - 0.5\theta\cos 0.5\theta}\right)^{-1}.$$

The stress effect on a single transistor PA shown in Fig. 4.1 is examined. In Fig. 4.1, the 1 nH inductor and 30 pF capacitor of the high-Q tank are used to short higher order harmonics at the fundamental frequency. The transistor gate is biased to a quiescent point so that the amplifier is operated in the class AB mode. The input drive level is adjusted that the drain current swings between zero and predetermined maximum current.

The conductance angle of the PA subjected to hot electron and temperature stress is simulated [2]. The conduction angle versus stress time for three different temperatures (320, 350, and 400 K) is displayed in Fig. 4.2. As seen in Fig. 4.2, the PA moves from class AB, through class B, to C mode of operation during hot electron stress. In Fig. 4.2, the class AB PA becomes a class B PA after 1800s of stress at 400 K and then enters into class C mode of operation when stress continues. The degradation of conduction angle with stress time is signified when the temperature increases due to increased threshold voltage at high temperature.

As seen in (4.1) and (4.2), the output power and efficiency are a strong function of conduction angle. It is anticipated that the output power and power efficiency decrease with stress time. The degradation is larger when temperature increases.

The electrical stress effect on a single transistor PA operating at 60 GHz was examined [3]. Simple analytical equation to relate the drift of saturation power with

Fig. 4.1 A single transistor power amplifier (© IEEE)

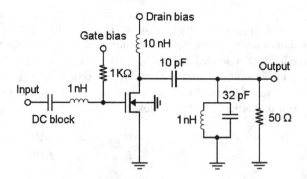

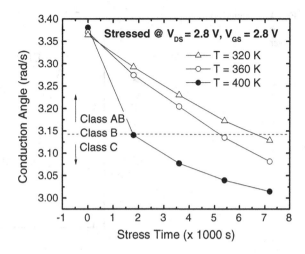

Fig. 4.2 Extracted conduction angle versus stress time (© IEEE)

device degradation has been identified. The normalized saturated output power can be expressed as

$$\frac{\Delta P_{\text{sat}}}{P_{\text{sat}}} \approx \frac{\Delta I_{ds}}{I_{ds}} \approx -\frac{2\Delta V_T}{V_T} + \frac{\Delta \mu_n}{\mu_n} \tag{4.5}$$

The above equation is validated through the measurement data. Those experimental results demonstrate that the hot-carrier stress reliability model is accurate at millimeter-wave frequencies to predict the degradation of the small- and large-signal PA performances. Experimental data also show that threshold voltage shift is the largest contributor to the deterioration of the PA performances. A decrease of 16 % of the power gain and 17 % of the 1 dB compression power are measured after a stress of 50 h of $V_{\text{DD}} = 1.65$ V at $P_{\text{in}} = 0$ dBm and $V_{\text{DD}} = 1.9$ V at $P_{\text{in}} = -10$ dBm at 60 GHz [3].

It is interesting to point out that the oxide breakdown effect may also be important for the class A, B, or C power amplifier shown in Fig. 4.1 due to significant drain-gate voltage stress at high RF input power, as evidenced by the simulation waveforms in Figs. 4.3 and 4.4. For example, at $t = 18.7$ ns the drain-gate voltage of the single transistor PA is 2.5 V at low RF input power (see Fig. 4.3). In this case, hot electron injection effect is likely. The drain-gate voltage increases to 3.5 V, however, when the RF input power level is high (see Fig. 4.4). High drain-gate stress may initiate the gate oxide breakdown effect, which degrades the transistor and circuit performance.

Fig. 4.3 Simulated voltage versus time at low input power

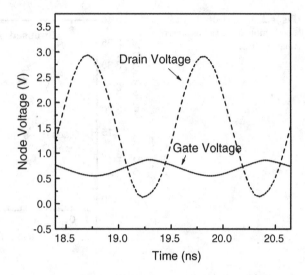

Fig. 4.4 Simulated voltage versus time at high input power

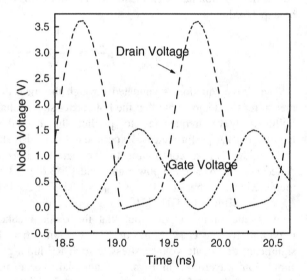

4.2 Class E Power Amplifier

Class E PA topology has become popular in wireless communication ICs due to its high power efficiency [4], and therefore a good candidate for low-cost, high integration portable communication systems such as cell phones, wireless local area networks, wireless sensor networks, global positioning systems, and Bluetooth applications. A class E amplifier with a shunt capacitor was introduced by Sokal and Sokal [5] and was examined by Rabb in an analysis of idealized operation [6].

The active transistor in the class E PA is used as a switch. The voltage waveform and current waveform are shaped by the LC tuning network such that they do not overlap, producing an ideal power efficiency of 100 %.

It is known that class E PA is very vulnerable to oxide stress because its drain voltage can approach more than three times of supply voltage V_{DD} ideally. To ensure the reliability of class E PA operation, V_{DD} is conservatively selected to a lower value. However, low V_{DD} reduces a PA's output power and power efficiency. To evaluate the class E PA reliability by experiments, a cascode class E PA is designed for fabrication [7]. Figure 4.5 shows the circuit schematics of a cascode class E PA. Using TSMC 0.18 μm mixed-signal CMOS technology, the class E PA designed at 5.2 GHz is evaluated in ADS simulation. Multifinger transistors with n-channel length of 0.18 μm are used. The driver transistor M_1 has the channel width of 256 μm. The main transistor M_2 and cascode transistor M_3 have the channel width of 512 and 544 μm, respectively. The DC supply voltage V_{DD1} for the driver stage is set at 1 V. The supply voltage V_{DD2} for the main amplifier is selected to be at 2.4 V. To reduce power consumption, the gate of M_1 is biased at 0.1 V (class C mode of biasing). The gate DC voltages of M_2 and M_3 are at 0.7 and 1 V, respectively. The spiral inductor and capacitor values used in this design are L_{in} = 3.61 nH, L_{D1} = 1.47 nH, L_{D2} = 4.56 nH, L_{tr} = 0.27 nH, L_S = 3.61 nH, C_{in} = 398 fF, C_{shunt} = 1.79 pF, C_{mid} = 1.68 pF, C_{tr} = 804 fF, C_S = 398 fF, and C_{OUT} = 35.6 fF. The feedback resistance $R_{feedback}$ is 600 Ω. The feedback transistor is used to improve the stability of the amplifier.

The cascode class E power amplifier was laid out using Cadence Virtuoso software [8], followed by successful Caliber DRC for design rule checking and LVS for layout versus schematic verification. The layout parasitic effects were extracted using ADS Momentum (EM) simulation [9]. Post layout simulation was performed to verify the design specification.

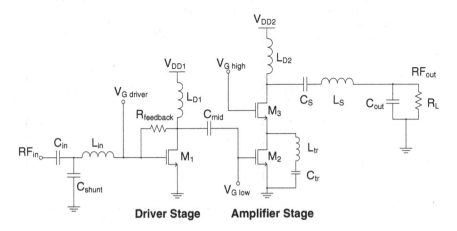

Fig. 4.5 Schematic of a cascode class E power amplifier (© IEEE)

The gate-source voltage and drain-source voltage as a function of time for the cascode transistor and main transistor are shown in Fig. 4.6 to examine the electrical stress effect on this cascode class E design. As seen in Fig. 4.6, the cascode transistor bears more voltage stress at the drain of *M3* than that of the main transistor *M2*. At high input power, the cascode transistor could suffer hot electron effect when gate-source voltage and drain-source voltage are high during switching transient (see Fig. 4.7).

Fig. 4.6 Simulated gate-source and drain-source voltage of main transistor. In this simulation $P_{in} = 0$ dBm and $V_{DD2} = 2.4$ V (© IEEE)

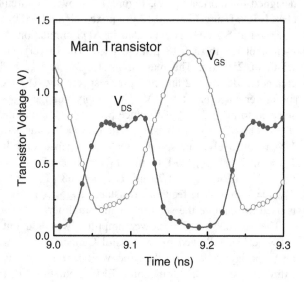

Fig. 4.7 Simulated gate-source and drain-source voltage of cascode transistor. In this simulation $P_{in} = 0$ dBm and $V_{DD2} = 2.4$ V (© IEEE)

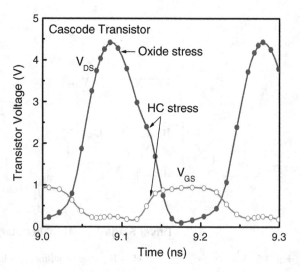

4.2.1   Mixed-Mode Device and Circuit Simulation

To evaluate the physical insight of hot electron effect in the cascode PA, the Sentaurus TCAD software is used [10] and the class E amplifier stage of the cascode power amplifier in Fig. 4.1 is simulated. It is worth pointing out that the mixed-mode device and circuit simulation provides the examination of device physical insight under the practical circuit operation condition. Figure 4.8 shows impact ionization rates for the cascode transistor and main transistor with increased supply voltage $V_{DD2} = 3.5$ V for accelerated aging. As seen in Fig. 4.8, the impact ionization rates of the cascode transistor are much higher than those of the main transistor due to higher electric field at the drain of the cascode transistor. Larger drain-source voltage also makes impact ionization rates at the peak of output voltage transient (left figure) higher than those during the output switching (right figure) as seen in Fig. 4.9. High impact ionization rates near the drain of MOS

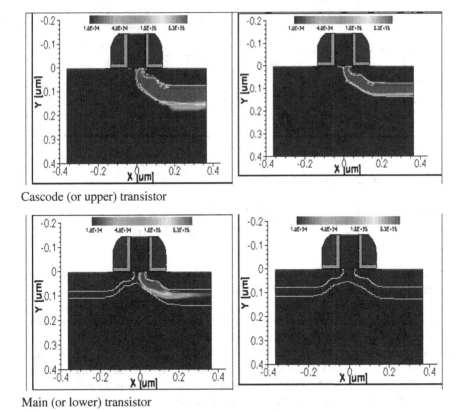

Cascode (or upper) transistor

Main (or lower) transistor

Fig. 4.8 Impact ionization rates of the cascode transistor (*upper plots*) and main transistor (*lower plots*) at the maximum (*left figures*) and middle (*right figures*) of the output voltage transient. In this mixed-mode device and circuit simulation, $V_{DD2} = 3.5$ V (© IEEE)

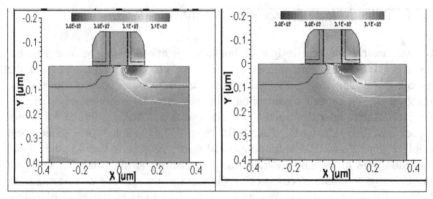

Cascode (or upper) transistor

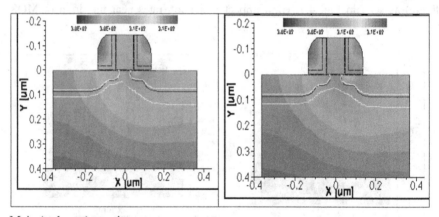

Main (or lower) transistor

Fig. 4.9 Lattice temperature of the cascode transistor (*upper plots*) and main transistor (*lower plots*) at the maximum (*left figures*) and middle (*right figures*) of the output voltage transient. In this mixed-mode device and circuit simulation, $V_{DD2} = 3.5$ V (© IEEE)

transistor ($\sim 6.3 \times 10^{26}/cm^3/s$) suggest that a large amount of hot electrons may inject into the gate of the MOSFET. Some hot electrons could be trapped within the oxide without reaching the gate contact. The accumulated trapped electron charges over a period of time increases the threshold voltage of the MOSFET. In addition, the interfacial layer between the SiO_2 and Si interface near the drain region may be damaged or degraded by the channel hot electrons. Thus, the effective electron mobility of the MOSFET decreases. Consequently, the drain current and transconductance of the MOSFET decrease. The reduction in drain current could decrease the output power and efficiency of the power amplifier as demonstrated by the experimental data in Figs. 4.11 and 4.12.

Figure 4.9 displays lattice temperature of the cascode transistor and main transistor. In Sentaurus simulation, Thermodynamic, Thermode, RecGenHeat, and

AnalyticTEP models are used to account for lattice heating. The thermodynamic model extends the drift-diffusion approach to account for electrothermal effects. A Thermode is a boundary where the Dirichlet boundary condition is set for the lattice. RecGenHeat includes generation–recombination heat sources. AnalyticTEP gives analytical expression for thermoelectric power. The substrate of the nMOS is set to be at 300 K. In Fig. 4.9, the cascode transistor has a higher peak lattice temperature ($\sim$310 K) than that in the main transistor because of larger drain-source voltage and power dissipation in the cascode transistor. The self-heating effect is enhanced during the output voltage switching (right figures in Fig. 4.9) because of relatively high drain-source voltage and high drain current simultaneously. High temperature rise resulting from lattice self-heating could further reduce drain current of the power amplifier. Consequently, the output power and power-added efficiency of the power amplifier degrade even more due to lattice heating. Note, it is well known that the class E power amplifier is vulnerable to the gate oxide breakdown due to very high drain-gate field stress. Here, however, we demonstrate the experimental data that the cascode class E power amplifier is degraded by hot electron effect during high output voltage switching. The mixed-mode device and circuit simulation of high impact ionization rates for the cascode transistor here supports the experimental finding of the PA degradation subjected to DC supply voltage for 10 h of continued RF stress at the input power of 0 dBm. The impact ionization leads to the formation of electron-hole pairs: electrons can be trapped in the gate oxide, whereas holes can generate interface states. Trapped electrons increase the threshold voltage of the n-channel MOSFET, while interface states may degrade the effective channel electron mobility. For the power amplifier performance degradation, threshold voltage shift is more important than mobility degradation [11]. Note, high input power RF stress could result in more degradation in hot electron effect than that under DC stress [12].

Additional mixed-mode simulation at RF stress under $V_{DD2} = 4.5$ V condition indicates that the peak impact ionization rates increase to $6.3 \times 10^{27}/\text{cm}^3/\text{s}$ and the maximum lattice temperature of the cascade transistor is about 320 K. This suggests that the hot electron effect and lattice heating are enhanced resulting from a higher drain electric field when V_{DD2} is at 4.5 V. High temperature from lattice heating, however, could reduce the hot electron effect compared to that without self-heating [12]. On the other hand, high temperature enhances gate oxide breakdown which is a strong function of temperature and electric field [13].

4.2.2 Chip Fabrication and Experimental Data

A silicon chip of the designed PA was fabricated using TSMC 0.18 μm mixed-signal CMOS technology [7]. The silicon chip is shown in Fig. 4.10 and its size is 820×887 μm^2. In this figure, spiral inductors, capacitors, transistors, GSG RF input pads and output pads, DC supply voltage pads, and gate bias pads are displayed. The PA's performances before and after RF stress are measured. The

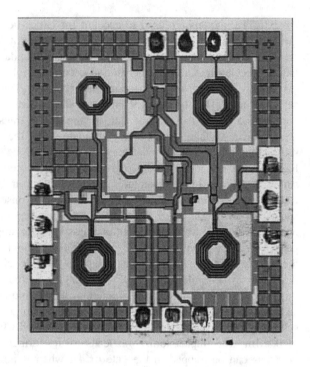

measurement was performed at room temperature. The PA was then stressed with
an RF input power of 0 dBm and different V_{DD2} stress level at 3.5, 4.0, or 4.5 V for
10 h. After continuous RF and elevated DC stresses, the RF parameters were
measured again to compare with the experimental data obtained at the fresh circuit
condition. For the circuit at the normal operation, DC biases of $V_{G1} = 0.1$ V,
$V_{G2} = 0.7$ V, $V_{DD1} = 1$ V, and $V_{DD2} = 2.4$ V were used.

Figure 4.11 shows the measured small-signal gain S_{21} versus frequency as a
function of stress conditions. In Fig. 4.11, the solid line represents the fresh circuit
result, the circles represent the PA's experimental data after 10 h of RF stress at
$V_{DD2} = 3.5$ V, the triangles represent the data after 10 h of RF stress at
$V_{DD2} = 4.0$ V, and the squares represent the measured result after 10 h of RF stress
at $V_{DD2} = 4.5$ V. As seen in Fig. 4.11, the larger the elevated stress at high V_{DD2},
the larger the S_{21} degradation over a wide range of frequencies. At 5.2 GHz, the
measured output power and power gain are plotted in Fig. 4.12. The output power
increases with input power and reaches a saturated output power at high input
power, thus making the power gain decrease at high input power. Both the output
power and power gain decrease after RF stress, especially when V_{DD2} stress level is
increased. The degradations of RF circuit performances are attributed to hot electron effect on the output transistor.

The measured power-added efficiency is illustrated in Fig. 4.13. Power-added
efficiency increases with input power, reaches saturation, and then decreases with
input power due to reduced output power and increased DC power dissipation at

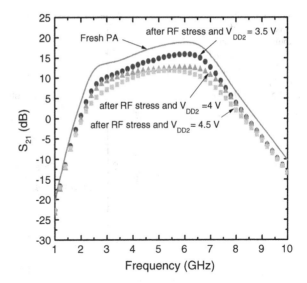

Fig. 4.11 Measured S_{21} versus frequency before and after RF stress. During the RF stress P_{in} is at 0 dBm and V_{DD2} was kept at 3.5, 4, or 4.5 V (© IEEE)

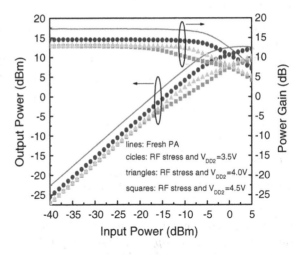

Fig. 4.12 Measured output power and power gain versus input power before and after RF stress. During this RF stress P_{in} is at 0 dBm and V_{DD2} was kept at 3.5, 4, or 4.5 V (© IEEE)

very high input power. The power-added efficiency is defined as (RF output power–RF input power)/total DC power dissipation including the power stage's and driver stage's. Note that the power-added efficiency is lower than the drain efficiency because of additional power dissipation from the driver stage. At 5.2 GHz, the peak power-added efficiency of the fresh PA approaches 25 % (a somewhat lower value than expected due to layout parasitic effect and additional DC power dissipation in the driver stage). After RF stress the peak power efficiency decreases significantly, especially when the V_{DD2} stress level is high.

Table 1 lists the small-signal gain S_{21} at 5.2 GHz, output power at the input power of 0 dBm, power gain (p_o/p_i) at the input power of −20 dBm, and maximum

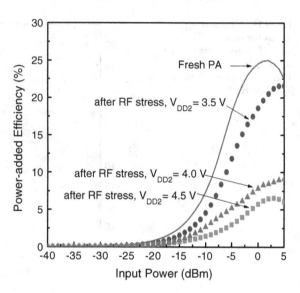

Fig. 4.13 Measured power-added efficiency versus input power before and after RF stress at 5.2 GHz. During this RF stress P_{in} is at 0 dBm and V_{DD2} was kept at 3.5, 4, or 4.5 V (© IEEE)

Table 1 RF parameter degradations

RF parameters	S_{21} at 5.2 GHz	p_o at $p_i = 0$ dBm	Gain at -20 dBm	Peak η_{add} (%)
Fresh	18.2 dB	12.5 dBm	17.3 dB	25
After RF stress[a]	15.2 dB	10.6 dBm	14.5 dB	21.6
After RF stress[b]	12.3 dB	7.9 dBm	12.9 dB	9.1
After RF stress[c]	11.9 dB	7.3 dBm	12.5 dB	6.6
Change[a]	-16.5 %	-15.2 %	-16.2 %	-13.6
Change[b]	-32.4 %	-36.8 %	-25.4 %	-63.6
Change[c]	-34.6 %	-41.6 %	-27.7 %	-73.6

[a]RF stress at $p_i = 0$ dBm and $V_{DD2} = 3.5$ V for 10 h
[b]RF stress at $p_i = 0$ dBm and $V_{DD2} = 4.0$ V for 10 h
[c]RF stress at $p_i = 0$ dBm and $V_{DD2} = 4.5$ V for 10 h

power-added efficiency before and after RF stress. Their normalized parameter shifts such as $\Delta S_{21}/S_{21}$, $\Delta p_o/p_o$, $\Delta(p_o/p_i)/(p_o/p_i)$, $\Delta\eta_{add}/\eta_{add} \times 100$ % from the fresh condition are also shown in Table 1.

In our stress experiments, no noticeable increase in gate leakage current was detected when V_{DD2} was stressed at 3.5, 4.0, and 4.5 V. This suggests no transistor oxide hard breakdown occurred since hard breakdown typically results in a sudden surge of gate current [14, 15], and could collapse RF performances. In addition, ADS circuit simulation indicates that the peak drain-gate voltage of the cascode transistor with the oxide thickness of 4.08 nm result in a smaller electric field than the critical field for oxide breakdown [16]. The oxide under this high RF and elevated DC stress at $V_{DD2} = 4.5$ V may experience some kind of soft breakdown which deteriorates the PA circuit performances further. Soft breakdown increases the gate leakage current noise due to formulation of random defects and conducting path within the oxide [17]. After soft breakdown, the nMOS transistor's threshold voltage is increased [18, 19] due to more trapped charge or defect density in the oxide. The increase in threshold voltage causes a decrease in drain current. Consequently, the PA's output power and power efficiency decrease after soft breakdown (to the first order, $\Delta P_o/P_o$ is proportional to $\Delta I_D/I_D$).

References

1. Bahl IJ (2009) Fundeamental of RF and microwave transistor amplifiers. Wiley, New York
2. Yu C, Yuan JS (2007) Electrical and temperature stress effects on class-AB power amplifier performances. IEEE Trans Electron Devices 54:1346–1350
3. Quemerais T, Moquillon L, Huard V, Fournier J-M, Benech P, Corrao N, Mescot X (2010) Hot-carrier stress effect on a CMOS 65-nm 60-GHz one-stage power amplifier. IEEE Electron Device Lett 927–929
4. Kalim D, Erguvan D, Negra R (2009) Broadband CMOS class-E power amplifier for LTE applications. Int Conf Sig Circ Syst 1–4
5. Sokal NO, Sokal AD (1975) Class E—A new class of high-efficiency tuned single-ended switching power amplifiers. IEEE J Solid-State Circuits 168–176
6. Raab FH (1977) Idealized operation of the class E tuned power amplifier. IEEE Trans Circuits Syst 725–735
7. Yuan JS, Yen HD, Chen SY, Wang RL, Huang GW, Juang YZ, Tu CH, Yeh WK, Ma J (2012) Experimental verification of RF stress effect on cascode class E PA performance and reliability. IEEE Trans Device Mater Reliab 369–375
8. http://www.cadence.com/products/cic/analog_design_environment/
9. http://www.agilent.com/find/eesof-ads
10. http://www.synopsys.com
11. Liu Y, Yuan JS (2011) CMOS RF power amplifier variability and reliability resilient biasing design and analysis. IEEE Trans Electron Devices 540–546
12. Quémerais T, Moquillon L, Huard V, Fournier J-M, Benech P, Corrao N, Mescot X (2010) Hot-carrier stress effect on a CMOS 65-nm 60-GHz one-stage power amplifier. IEEE Electron Device Lett 927–929
13. Kimura M (1999) Field and temperature acceleration model for time-dependent dielectric breakdown. IEEE Trans Electron Devices 220–229
14. Pompl T, Wurzer H, Kerber M, Eisele I (2000) Investigation of ultra-thin gate oxide reliability behavior by separate characterization of soft breakdown and hard breakdown. In: Technical Digital International Physics Symposium 2000, pp 40–47
15. Pantisano L, Cheung KP (2001) The impact of postbreakdown gate leakage on MOSFET RF performances. IEEE Electron Device Lett 585–587
16. Liu C-H, Wang R-L, Su Y-K (2010) DC and RF degradation induced by high RF power stresses in 0.18-µm nMOSFETs. IEEE Trans Device Mater Reliab 317–323
17. Roussel P, Degraeve R, van den Bosch C, Kaczer B, Groeseneken G (2001) Accurate and robust noise-based trigger algorithm for soft breakdown detection in ultra thin oxides. In: Technical Digital International Reliability Physics Symposium 2001, pp 386–392
18. Avellán A, Krautschneider WH (2004) Impact of soft and hard breakdown on analog and digital circuits. IEEE Trans Device Mater Reliab 676–680
19. Rodríguez R, Stathis JH, Liner BP (2003) A model for gate-oxide breakdown in CMOS inverters. IEEE Electron Device Lett 114–116

Chapter 5
Voltage-Controlled Oscillator Reliability

In the wireless communication ICs, the voltage-controlled oscillator is used for both transmitter and receiver circuits. A great deal of research in the past two decades has focused on integrated VCOs using transformer feedback [1, 2], injection lock [3, 4], and current reuse [2, 4–7], which were made possible with the advancement of CMOS technology. When evaluating performances of VCOs, several parameters such as phase noise, oscillation frequency, tuning range, harmonic output power, and power consumption need to be considered.

5.1 Differential-Mode LC Oscillator

The voltage stress affects the oscillator performance such as oscillator frequency and phase noise. To evaluate the reliability issues by experiments, an LC oscillator is designed and fabricated. Figure 5.1 shows the schematic of the LC-VCO. The inductors $L1$ and $L2$ and the varactors C_{V1} and C_{V2} form the basic LC resonator. The accumulation-mode MOS varactors C_{V1} and C_{V2} are used for frequency tuning. The voltage V_{tune} is used to tune the capacitance of varactors and the oscillation frequency. The VCO can also be designed with only the inductors $L1$ without the inductor $L2$ with smaller sized inductor. The cross-coupled pair ($M1$ and $M2$) generates a negative transconductance to compensate for the LC tank loss. The n-core rather than p-core VCO topology is chosen since the nMOS has larger transconductance gain than pMOS. A key to achieving wide-tuning range is minimizing the parasitic capacitances connected to the tank. MOSFETs with a smaller threshold voltage are used to reduce the dimension of MOSFETs while maintaining enough transconductance gain to start the oscillation. A large varactor is used in the design, the parasitic capacitance of active transistors is much smaller than the

© The Author(s) 2016
J.-S. Yuan, *CMOS RF Circuit Design for Reliability and Variability*,
SpringerBriefs in Reliability, DOI 10.1007/978-981-10-0884-9_5

Fig. 5.1 Schematic of the differential LC-VCO

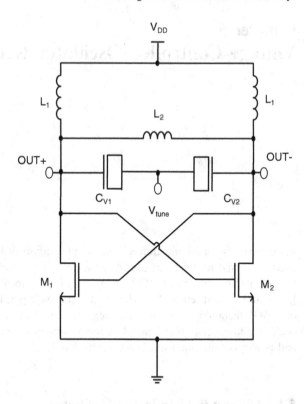

capacitance of varactors, and the tuning range is mainly determined by the property of variable capacitors. Therefore, the VCO performance is less sensitive to process variation.

The VCO was laid out using Cadence Virtuoso software [8], followed by Caliber DRC for design rule checking and LVS for layout versus schematic verification. The VCO has been implemented in the retrograde twin-well UMC 90-nm 1P9M standard CMOS technology. The fabricated die size is 0.71×0.48 mm^2. Its micrograph is depicted in Fig. 5.2. In this figure, spiral inductors, capacitors, transistors, GSG RF output pads, DC supply voltage pads, and gate bias pads are shown.

To evaluate the physical insight of hot electron effect in RF operation, the mixed-mode simulation of Sentaurus TCAD software is used [9]. The mixed-mode device and circuit simulation allows one to evaluate the device physical insight under the oscillator large-signal operation condition. In Sentaurus simulation, physical equations such as Poisson's and continuity equations for drift-diffusion transport are implemented. The Shockley-Read-Hall carrier recombination, Auger recombination, and impact ionization models are also used.

Fig. 5.2 Microphotograph of the measured VCO

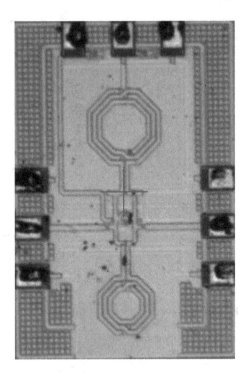

5.1.1 Circuit and Device Simulation

To identify the proper overstress condition for the VCO, ADS circuit simulation is used. Figure 5.3 shows the gate-source voltage and drain-source voltage versus time. This plot demonstrates that the MOSFET is under hot electron stress and off-state avalanche. Figure 5.4 shows impact ionization rates for the n-channel transistor in the oscillator circuit as shown in Fig. 5.1. The circuit element values used in the mixed-mode device and circuit simulation are the same as those designed in Fig. 5.1. The supply voltage in Sentaurus simulation is set at $V_{DD} = 1.5$ V to mimic the accelerated aging condition suggested by Fig. 5.3. The impact ionization rates during the switching transient (on-state) and near the peak of output voltage (off-state) and their corresponding electric field distributions are shown in Figs. 5.4 and 5.5, respectively. High impact ionization rates (close to $10^{24}/$ cm^3/s) at the drain of MOS transistor indicate hot electron effect near the drain edge during switching transition and avalanching at the off-state (near the peak of V_{DS}). During avalanche a strong lateral electric field exists in the device leading to collision of electrons with atomic bonds at Si–SiO$_2$ interface and breaking them and creating dangling bonds, or surface states.

Fig. 5.3 Output waveforms versus time for accelerated hot electron stress

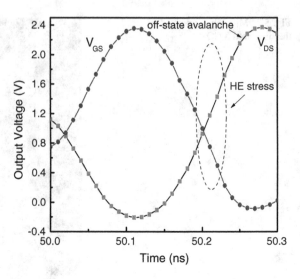

Fig. 5.4 a Impact ionzation rates during the switching transient, **b** impact ionzation rates at off-state

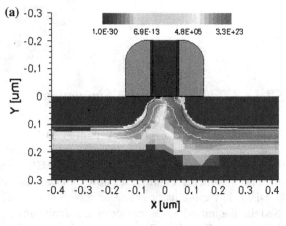

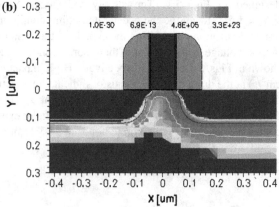

Fig. 5.5 **a** Electric field during the switching transient, **b** electric field at off-state

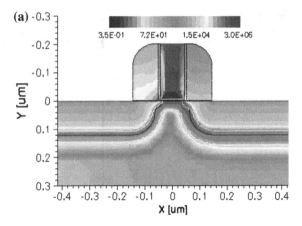

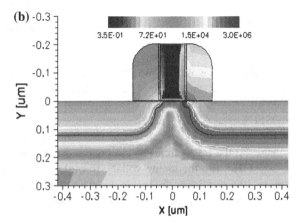

5.1.2 Experimental Data

The oscillator's performances before and after RF stress are measured. The experiment was performed at room temperature with an Agilent E5052 Signal Source Analyzer. During the measurement, one of the oscillator's outputs is connected to the Signal Source Analyzer, while the other output is connected to a 50 Ω load. The LC-VCO was then stressed under the continued RF oscillation (~8.5 GHz) condition at $V_{tune} = 0.35$ V by increasing the supply voltage V_{DD} from 1 to 1.5 V for accelerated aging. After RF oscillation stress at $V_{DD} = 1.5$ V, the RF parameters were measured at 1 and 5 h time points. The measurement was done at the normalized circuit operation condition of $V_{DD} = 1$ V.

Figure 5.6 shows the measured VCO core current versus tuning voltage. The current of the cross-coupled VCO is about 18.9 mA under 1 V supply voltage at the fresh circuit condition. After RF stress, the VCO current decreases. For example, the current is 17.2 mA after 1 h of RF stress and 15.6 mA after 5 h of RF stress, at

Fig. 5.6 Measured VCO core
current versus tuning voltage
before and after RF stress

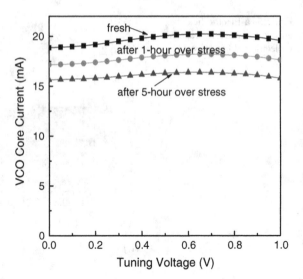

the tuning voltage of 0 V. The decrease in VCO current is attributed to the reduction
of drain current due to hot electron effect and off-state avalanche. Hot electron
injection increases the threshold voltage and thus decreases the drain current of the
MOSFET.

Figure 5.7 shows the measured oscillation frequency as a function of tuning
voltage. At the fresh circuit condition, the oscillation frequency changes from 8.65
to 10.2 GHz with the tuning voltage from 0 to 1.0 V. After RF and increased V_{DD}
stresses, the oscillation frequency increases. For example, at the tuning voltage of
0.35 V, the oscillation frequency increases from 8.92 GHz before stress to

Fig. 5.7 Measured
oscillation frequency versus
tuning voltage before and
after RF stress

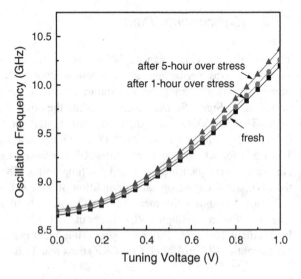

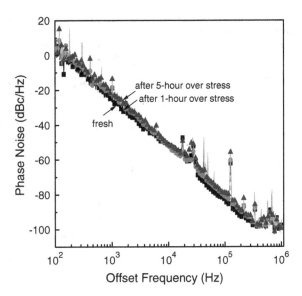

Fig. 5.8 Measured phase noise versus offset frequency before and after RF stress ($V_{tune} = 0.35$ V)

8.98 GHz after 5 h of stress. Hot electron stress decreases the gate capacitance of SiO_2 MOSFETs [10]. The decrease in total capacitance (including the capacitances from the active transistors and varactors) of the resonant increases the frequency of oscillation.

The phase noise as a function of offset frequency from the carrier is displayed in Fig. 5.8. The phase noise decreases with the offset frequency, as expected. It can be seen from Fig. 5.8 that the measured phase noise at low offset frequency has a slope of −30 dBc/dec, which indicates that $1/f$ noise is up-converted to the carrier phase noise. Clearly, the phase noise increases after hot electron stress for 5 h. A zoom-in plot from 1 to 10 kHz is displayed in Fig. 5.9 for better visualization of the phase noise difference between the fresh and stressed data. The increase in phase noise after hot electron effect could be related to decreased oscillator's output amplitude and increased interface states which contribute flicker noise. For example, the $1/(\Delta f)^3$ dependence in the phase noise spectrum observed in Figs. 5.8 and 5.9 at low offset frequency is due to $1/f$ noise up-conversion to the carrier phase noise. Hot carrier damage is known to significantly increase the $1/f$ noise of the nMOSFET. Flicker noise is produced by traps at the SiO_2 and silicon interface. Hot carrier induced traps are known to produce the same kind of noise from traps at the SiO_2 and Si interface [11]. These interface traps can capture or reemit charges from or to the channel in a random process with relatively long time constants. The resulting fluctuations in the drain current mainfest themselves as low frequency noise with $1/f$ frequency dependence. In addition, the low frequency noise under periodic large-signal excitation (such as in the LC-VCO) increases more rapidly due to hot carrier degradation as compared to the low frequency noise measured under steady state [12]. It is worthy pointing out that off-state drain stress emerges as a significant degradation mechanism for large-signal RF applications. Its degradation is

Fig. 5.9 Phase noise plot from 1 to 10 kHz offset frequency ($V_{\text{tune}} = 0.35$ V)

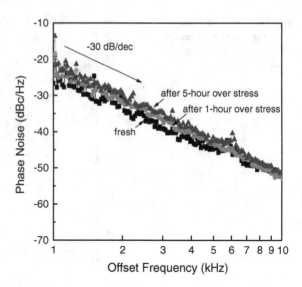

comparable to that of the channel hot electron effect [13]. The normalized drain current flicker noise is correlated with the associated shift of drain current in the linear region ($-\Delta I_{\text{dlin}}$) of the nMOS LC oscillator [14]. These reliability effects in turn increase the phase noise of the oscillator.

The oscillator's figure of merit (FOM) is usually defined as:

$$\text{FOM} \equiv L(\Delta f) + 10 \times \log\left(\frac{P_{\text{DC}}}{1\,\text{mW}}\right) - 20 \times \log\left(\frac{f_0}{\Delta f}\right) \qquad (5.1)$$

where L is the phase noise, P_{DC} is the DC power dissipation, f_o is the oscillation frequency, and Δf is the offset frequency. The FOM before and after 1 and 5 h of RF stress is given in the following Table (@$V_{\text{tune}} = 0.35$ V, $\Delta f = 100$ kHz, and $V_{\text{DD}} = 1.0$ V):

It is clear from Table 5.1 that the FOM degrades from -169.29 dBc/Hz (fresh condition) to -164.86 dBc/Hz (after 5 h of RF stress) at an offset frequency of 100 kHz.

Table 5.1 LC oscillator performance before and after stress

	$L(\Delta f)$ (dBc/Hz)	$10 \times \log(P_{\text{DC}}/1\,\text{mW})$	$-20 \times \log(f_0/\Delta f)$	FOM (dBc/Hz)
Fresh	-83.22	12.94	-99.00	-169.29
1 h stress	-80.67	12.49	-99.04	-167.22
5 h stress	-77.86	12.06	-99.06	-164.86

5.2 LC Loaded (or Current Reuse) VCO

For low-power application, the current reuse topology is preferred. Figure 5.10 shows the circuit schematics of a current reused low-power oscillator. This VCO has a p-channel MOSFET M1, an n-channel MOSFET M2, two balanced resistors, one inductor, and two p-channel transistor varator diodes in the LC tank, and two output buffers at the differential outputs. The current reused LC-VCO replaces one of the n-channel MOSFET of a conventional differential LC-VCO with a p-channel MOSFET. Both n-channel and p-channel transistors in the cross-connected pair are a negative conductance generator. Unlike a conventional VCO where the transistors switch alternately, the current reused VCO does not have a common-source node because the transistors switch on and off at the same time. The current reused VCOs can operate with only half the amount of DC current compared to those of the conventional VCO topologies [15]. For instance, during the first half period, both n- and p-channel transistors are on and the current flows from V_{DD} to ground through the tuning inductor. During the second half period, the two transistors are off and the current flows in the opposite direction through the capacitors. The output buffers are used to improve the pull effect resulted from various load. pMOS transistors are used as varactors due to their lower $1/f$ noise. Note that typical LC oscillator design [16] does not require the use of resistors. In our design, however, two resistors are added to improve magnitude asymmetry of differential outputs, which is resulted from different transconductances of two cross-connected transistors or different equivalent impedances at two drain terminals, at the expense of minor increase of power dissipation due to resistors.

Fig. 5.10 Schematic of current reused VCO with balanced resistors and buffers (© Elsevier)

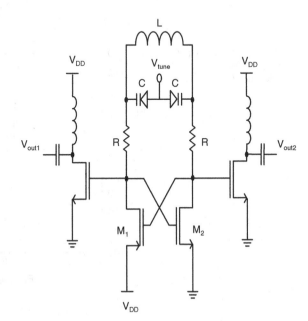

The current reused VCO was laid out using Cadence Virtuoso software, followed by successful Calibre DRC for design rule checking and LVS for layout versus schematic verification. ADS simulation results of oscillating waveforms are displayed in Fig. 5.2a. The output waveforms are differential and the oscillation frequency estimated from the output waveforms is about 2.56 GHz. The output power spectrum in Fig. 5.2b shows that the fundamental power is −5.27 dBm and its second harmonic power is −20.5 dBm (Figs. 5.11 and 5.12).

Fig. 5.11 Simulated output voltage waveforms versus time (© Elsevier)

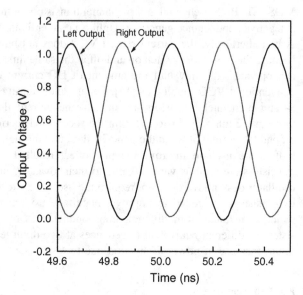

Fig. 5.12 Simulated output power spectrum characteristics (© Elsevier)

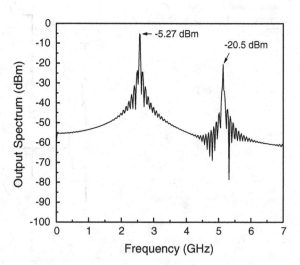

5.2.1   Fabrication and Experimental Data

The LC oscillator designed above fabricated using TSMC 0.18 μm mixed-signal CMOS technology. The n-channel and p-channel MOSFETs have 0.18 μm channel length and channel width of 1.5 μm/finger × 32 fingers. The p^+ polysilicon resistor R is 10 Ω and the RF spiral inductor L in the core is 8.74 nH. The fabricated silicon chip is shown in Fig. 5.13 and its size is 1.2 mm × 1.1 mm. In Fig. 5.13 spiral inductors, capacitors, transistors, GSG RF output pads, DC supply voltage pads, and gate bias pads are displayed.

The oscillator's performances before and after RF stress are measured. The measurement was performed at room temperature with an Agilent E5052 Signal Source Analyzer. The LC-VCO was then stressed under the continued RF large-signal transient condition with an increased supply voltage of 2.4 V for fast aging (for the normal operation, the supply voltage is at 1.2 V). After RF and increased DC stresses, the RF parameters were measured at 5 and 10 h time points. The measurement was done at the normalized circuit operation condition of $V_{DD} = 1.2$ V. To increase voltage stress, the supply voltage V_{DD} is set at 2.4 V to elevate the drain-source voltage and gate-source voltage during switching. This will enhance hot carrier and NBTI effect.

Figure 5.14 shows the gate-source voltage and drain-source voltage versus time from ADS circuit simulation. This plot demonstrates that the MOSFET is under hot

Fig. 5.13 Microphotograph of the fabricated current reused VCO (© Elsevier)

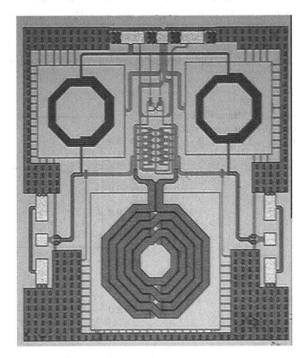

Fig. 5.14 Output waveforms
versus time for accelerated
hot electron stress,
V_{DD} = 2.4 V (© Elsevier)

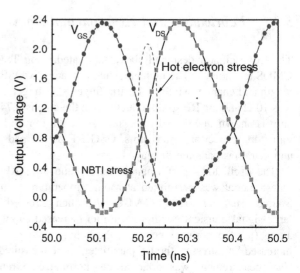

electron stress during transistor switching and the p-channel transistor is under NBTI stress prior to hot electron effect. For example, at 50.11 ns the gate-source voltage of the pMOS is less than −2.6 V (=−0.2 − V_{DD}), while its drain-source voltage is close to 0 V ($\approx$2.4 − V_{DD}, V_{DD} = 2.4 V). Figure 5.15 shows the measured output current versus tuning voltage. The total current of the current reused VCO is about 3.13 mA under 1.2 V supply voltage at the fresh circuit condition. Its measured VCO's output power at 2.4 GHz fundamental frequency is −5.165 dBm. After RF stress, the output current decreases. For example, the output current is 2.52 mA after 5 h of RF stress and 2.47 mA after 10 h of RF stress, at the tuning voltage of 0 V. The decrease in output current is attributed to the reduction of drain

Fig. 5.15 Measured output
current versus tuning voltage
before and after RF stress
(© Elsevier)

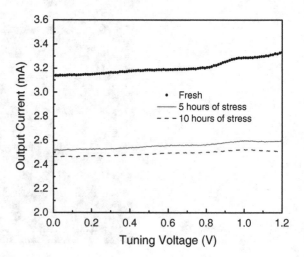

Fig. 5.16 Measured oscillation frequency versus tuning voltage before/after RF stress (© Elsevier)

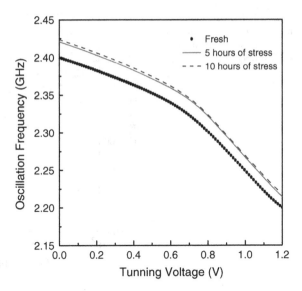

current due to hot electron effect and NBTI degradation. Hot electron injection increases the threshold voltage and decreases the channel mobility of the MOSFET, while the NBTI stress increases the magnitude of p-channel transistor threshold voltage. These effects in turns decrease the drain current flowing through the n-channel and p-channel transistors.

Figure 5.16 shows the measured oscillation frequency as a function of tuning voltage for the current reused VCO [17]. At the fresh circuit condition, the oscillation frequency changes from 2.4 to 2.2 GHz with the tuning voltage from 0 to 1.2 V. The oscillation frequency measured is slightly lower than the simulated value due to layout parasitic, interconnections, and bonding pad effects. The frequency range measured is suitable for wireless communications services. After RF and increased V_{DD} stresses, the oscillation frequency increases. For example, the oscillation frequency is 2.42 GHz after 5 h of stress and becomes 2.424 GHz after 10 h of stress, at the tuning voltage of 0 V. Hot electron stress decreases the gate-drain capacitance and increases the gate-source capacitance of SiO_2 MOSFETs [18, 19]. The decrease in gate-drain capacitance is larger than the increase in gate-source capacitance. Thus, the total capacitance of the resonant circuit decreases, which in turn increases the frequency of oscillation.

The phase noise as a function of offset frequency from the carrier is displayed in Fig. 5.17. The phase noise decreases with the offset frequency, as expected. At 1, 10, 100, and 1000 kHz offset frequency, the measured phase noise is −33.86, −63.38, −91.55, and −109.34 dBc/Hz, respectively, for the fresh circuit condition. The prestress measurement data are very close to the predicted phase noise values from simulation results. It can be seen from Fig. 5.17 that the measured phase noise at low offset frequency has a negative slope of 30 dBc/dec, indicating $1/f$ noise is up-converted with other white noise from different noise sources in the circuit.

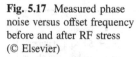

Fig. 5.17 Measured phase
noise versus offset frequency
before and after RF stress
(© Elsevier)

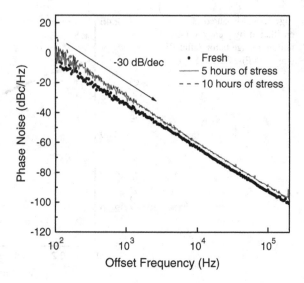

Furthermore, broad similarity between NBTI relaxation and $1/f$ noise was observed [20, 21]. Both NBTI relaxation and $1/f$ noise phenomena are caused by interface states with widely distributed time scales. This NBTI-related $1/f$ noise could contribute part of the measured phase noise in Fig. 5.17.

5.3 Dual-Band VCO

The explosion of portable wireless communication services results in the demand of multiband and multimode transceivers. Dual-band VCOs play an important role in dual-band transceivers required for multifunction services. The stress effect on dual-resonance VCO was studied [22]. The VCOs were designed and fabricated in the TSMC 0.18 μm 1P6 M CMOS technology. At $V_{tune2} = 0$ V, the dual-band VCO operates between 7.86–8.28 GHz at high band and 3.16–3.58 GHz at low band. The VCO transits from its first high parallel resonance frequency to its second low resonance frequency around V_{tune1} between 0.6 and 0.7 V. The dual-resonance VCO was stressed at $V_{DD} = 3$ V, $V_{tune1} = 0$ V, and $V_{tune2} = 0$ V for 1.5 and 3 h. After each stress period, the VCO property is measured again at different lower V_{DD}. High supply voltage increases the gate and drain voltage swings of switching transistors, and hot carrier injection occurs while the output voltage swing is high and the channel electric field is high. When a conventional NMOS is DC biased at maximum substrate current ($V_{GS} \cong 0.5V_{DS}$), the hot carrier damage is associated with the creation of interface states [23]. When the nMOSFET is biased at maximum gate bias (i.e., $V_{GS} = V_{DS}$, condition for hot electron injection), electron traps are created. At low V_{GS} (conditions for hot-hole injection), neutral electron traps are created. During the high V_{DD} stress carried out in the dual-band VCO, the VCO experiences different damage mechanisms [24]. The final dominant mechanism will shows its effect on the device characteristics and circuit performance. The hot carrier damage is localized while the circuit performance is a lumped effect of the overall nonuniform device damage. The dual-resonance VCO is sensitive to hot carrier stress as the hot carrier stress may cause the frequency band shifting out of the original design specification. The stressing also increases device current noises and leads to the increase in the phase noises at both high band and low band.

References

1. Ng WL, Luong HC (2007) A 1-V 17-GHz 5-mW CMOS quadrature VCO based on transformer coupling. IEEE J Solid State Circ 42:1933–1941
2. Huang T-H, Tseng Y-R (2008) A 1 V 2.2 mW 7 GHz CMOS quadrature VCO using current-reuse and cross-coupled transformer-feedback technology. IEEE Microwave Wireless Compon Lett 18:698–700
3. Chuang YH, Lee S-H, Yen R-H, Jang S-L, Lee J-F, Juang M-H (2006) A wide locking range and low voltage CMOS direct injection-locked frequency divider. IEEE Microwave Wireless Compon Lett 16:299–301
4. Lee S-H, Jang S-L, Chuang Y-H, Chao J-J, Lee J-F, Jung M-H (2007) A low power injection locked LC-tank oscillator with current reused topology. IEEE Microwave Wireless Compon Lett 17:220–222
5. Oh N-J, Lee S-G (2005) Current reused LC VCOs. IEEE Microwave Wireless Compon Lett 15:736–738
6. Yun S-Y, Shin S-B, Choi H-C, Lee S-G (2005) A 1 mW current-reuse CMOS differential LC-VCO with low phase noise. IEEE Int Solid State Circ Conf 540–541
7. Chuang Y-H, Jang S-L, Lee S-H, Yen R-H, Jhao J-J (2007) 5-GHz low power current-reused balanced CMOS differential Armstrong VCOs. IEEE Microwave Wireless Compon Lett 17:139–141, 299–301
8. http://www.cadence.com/products/cic/analog_design_environment/
9. http://www.synopsys.com
10. Ling CH, Ang DS, Tan SE (1995) Effects of measurement frequency and temperature anneal on differential gate capacitance spectra observed in hot carrier stressed MOSFETs. IEEE Electron Devices 5:1528–1535
11. Jin Z, Cressler JD, Abadeer W (2004) Hot-carrier stress induced low-frequency noise degradation in 0.13 μm and 0.18 μm RF CMOS technologies. In: International reliability physics symposium, pp 440–444
12. Brederlow R, Weberm W, Scgnjutt-Landsiedel D, Thewes R (2002) Hot-carrier degradation of the low-frequency noise in MOS transistors under analog and RF operating conditions. IEEE Trans Electron Devices 49:1588–1596
13. Kolhatkar J, Hoekstra E, Hof A, Salm C, Schmitz J, Wallinga H (2005) Impact of hot-carrier degradation on the low-frequency noise in MOSFETs under steady-state and periodic large-signal excitation. IEEE Electron Devices Lett 26:764–766
14. Reddy V, Barton N, Martin S, Hung CM, Krishnan A, Chancellor C, Sundar S, Tsao A, Corum D, Yanduru N, Madhavi S, Akhtar S, Pathak N, Srinivasan P, Shichijo S, Benaissa K, Roy A, Chatterjee T, Taylor R, Krick J, Brighton J, Ondrusek J, Barry D, Krishnan S (2009) Impact of transistor reliability on RF oscillator phase noise degradation. In: Proceedings international Electron Devices Meeting, pp 401–404
15. Yun S-Y, Shin S-B, Choi H-C, Lee S-G (2005) A 1 mW current-reuse CMOS differential LC-VCO with low phase noise. IEEE Int Solid State Circ Conf 540–541

16. Liu S-L, Chen K-H, Chang T, Chin A (2010) A low-power K-band CMOS VCO with four-coil transformer feedback. IEEE Microwave Wireless Compon Lett 20:459–461
17. Yen HD, Yuan JS, Wang RL, Huang GW, Yeh WK, Huang FS (2012) RF stress effects on CMOS LC-loaded VCO reliability evaluated by experiments. Microelectron Reliab 52: 2655–2659
18. Yu C, Yuan JS (2005) MOS RF reliability subject to dynamic voltage stress—modeling and analysis. IEEE Trans Electron Devices 52:1751–1758
19. Ling CH, Ang DS, Tan SE (1995) Effects of measurement frequency and temperature anneal on differential gate capacitance spectra observed in hot carrier stressed MOSFETs. IEEE Electron Devices 42:1528–1535
20. Luo Z, Walko JP (2005) Physical mechanism of NBTI relaxation by RF and noise performance of RF CMOS devices. In: International reliability physics symposium, pp 712–713
21. Kaczer B, Grasser T, Martin-Martinez J, Aoulaiche SM, Roussel PJ, Groseneken G (2009) NBTI from the perspective of defect states with widely distributed time scales. In: International reliability physics symposium, pp 55–60
22. Jang S-L, Sanjeev J, Huang J-F (2013) Experimental RF characteristics of hot-carrier-stressed p-core dual-band VCO. J Low Power Electron 9:247–252
23. Hu C, Tam SC, Hsu F-C, Ko P-K, Chan T-Y, Terrill KW (1985) Hot electron induced MOSFET degradation—model, monitor and improvement. IEEE J Solid State Circ 32: 295–305
24. Mistry KR, Doyle B (1993) AC versus DC hot-carrier degradation in n-channel MOSFETs. IEEE Trans. Electron Devices 40:96–104

Chapter 6
Mixer Reliability

Recently, active switching mixer design is widely used in CMOS millimeter-wave circuit design. Increasing demands for high speed and low cost home entertainment, imaging, and automotive radar systems lead to CMOS circuit and system research in millimeter-wave application. Some applications, such as HDMI wireless video receivers and point-to-point radio strongly require CMOS millimeter-wave devices to further implement SoCs [1]. The Gilbert cell structure has better isolation, conversion gain (CG), linearity, and compact size [2]. A double balanced Gilbert cell structure with Marchand baluns for broadband using TSMC 65 nm 1P6 M CMOS process was fabricated for the evaluation of mixer reliability. The Marchand balun provides simple structure, low amplitude/phase imbalance, and wideband frequency response [3, 4]. The stacked coupled balun is used for broadband matching and balance converting.

Figure 6.1 shows the simplified schematic of the double balanced mixer that consists of Marchand baluns for RF/LO input to obtain a wider range of bandwidth matching. The Marchand balun uses stacked coupled structure with top metal layer metal 6 ($M6$) and to metal 4 ($M4$). The response frequency is dominated by the length of the transmission line [5, 6]. The length of the transmission line is $\sim 1056\ \mu m$ and the line width is 12 μm for $M4$ and 3 μm for $M6$. It has a wide matching bandwidth from 60 to 86 GHz for the RF port and 58.8–83.2 GHz for LO port. In our mixer design, the RF and LO ports use the transmission line matching network. C_1, C_2, $TL_{11,12}$, $TL_{21,22}$ are the matching network for the internal stage of transconductance stage (M_1 and M_2), while C_3, C_4, $TL_{31,32}$, $TL_{41,42}$ are for the switching stage ($M_{3,4}$ and $M_{5,6}$). TL_{51} and TL_{52} are transmission lines of the width 3 μm and length 100 μm for source degeneration to enhance internal stage matching bandwidth [7]. TL_{61} and TL_{62} are matching components between the transconductance and switching stages to improve the conversion gain of the mixer. R_1 and R_2 are resistive loads and equal to 175 Ω for optimal conversion gain. The conversion gain of the mixer is designed to work with 7 dBm LO power at 68 GHz. The LO input power is between 4–7 dBm from 64 to 75 GHz. Since the balun insertion loss is about 6 dB, the actual power to the LO transistors reduces to −2 to 2 dBm.

© The Author(s) 2016
J.-S. Yuan, *CMOS RF Circuit Design for Reliability and Variability*,
SpringerBriefs in Reliability, DOI 10.1007/978-981-10-0884-9_6

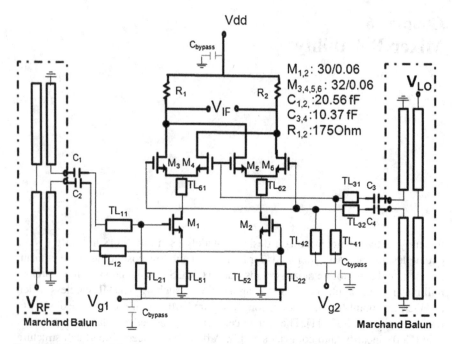

Fig. 6.1 Schematic of the down-converter mixer using Marchand baluns at RF/LO inputs

Fig. 6.2 Photograph of the down-conversion mixer

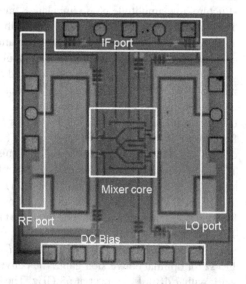

Figure 6.2 shows the die photograph of the fabricated millimeter-wave mixer using 65 nm 1P6 M CMOS technology. The chip size is 0.606 mm^2. A RF port (GSG) pad is on the left side and LO port (GSG) on the right. The top side is

balanced IF output (GS⁺GS⁻G). The DC pad is arranged in the order of V_{DD}, Gnd, V_{G1}, V_{G2}, Gnd, and V_{G2}.

Figure 6.3 shows measured input reflection coefficient of RF and LO ports at room temperature. The input stages of RF and LO transistors have a wide bandwidth matching from 60 to 86 GHz and 58.8–83.2 GHz for −10 dB reflection coefficient. The DC gate biasing for this measurement is under $V_{G1} = 0.7$ V and $V_{G2} = 0.8$ V. Figure 6.4 displays the measured isolation of the double balanced mixer LO-to-RF, LO-to-IF, and RF-to-IF from 60 to 85 GHz, where the isolation is less than −30 dB. The conversion gain performance of 68 GHz down-convert to f_{IF} at 1 GHz. Figure 6.5 shows the conversion gain versus the RF input power and the IF output power versus the LO input power. In Fig. 6.5, the solid squares represent

Fig. 6.3 Input reflection coefficient of input matching characteristics for RF and LO port

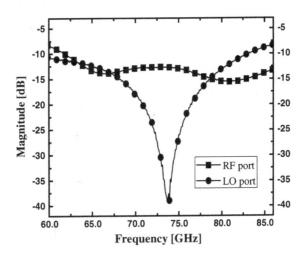

Fig. 6.4 Measured LO-to-RF, LO-to-IF, and RF-to-IF isolation verse RF frequency

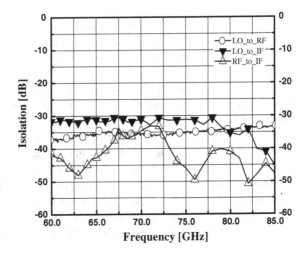

Fig. 6.5 Measured
conversion gain versus
RF/LO input power to
illustrate input compression
point (P1 dB) with RF of
68 GHz, a fixed IF frequency
of 1 GHz

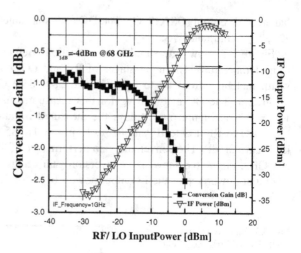

Fig. 6.6 Experimental data to
the current from V_{DD}
(the overstress voltage is
@ V_{DD} = 2.5 V)

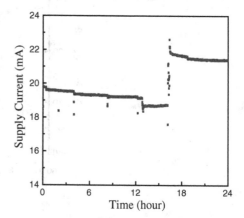

the conversion gain and the inverse triangle symbols represent the IF output power. As seen in Fig. 6.5, the saturation conversion gain is −0.93 dB. The corresponding IF output power at the 1 dB compression point (IP$_{1dB}$) is −4 dBm at 68 GHz.

RF stress experiment was performed under an elevated DC supply voltage V_{DD} of 2.5 V at room temperature. Figure 6.6 shows the measured mixer supply current from V_{DD} versus time. As seen in Fig. 6.6, the mixer supply current decreases with time in the beginning and then increases sharply, indicating oxide breakdown after about 17 h of dynamic stress at millimeter-wave frequency. The decrease in supply current in early 17 h is due to hot electron degradation that increases the threshold voltage and decreases the drain current of n-channel transistors of the mixer.

Figure 6.7 shows the mixer conversion gain versus RF frequency after different stress times. The conversion gain increases with RF frequency, but decreases with stress times, as expected. After 24 h of accelerated stress, the conversion gain can drop about 4 dB due to hot electron and oxide breakdown stress effects. At high RF

Fig. 6.7 Conversion gain versus RF frequency after different stress times

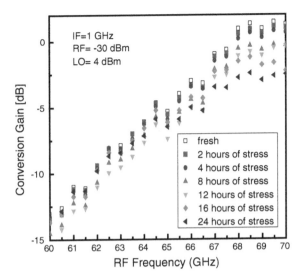

input power level, the conversion gain decreases with RF input power. The conversion gain also decreases with stress times at high RF input power at high RF input power, as observed by the measurement data in Fig. 6.7.

References

1. Cabric D, Chen M, Sobel D, Yang J, Broderson RW (2005) Future wireless systems: UWB, 60 GHz and cognitive radios. In: IEEE Custom Integrated Circuits Conference, 2005, pp 793–796
2. Mitomo T, Ono N, Hoshino H, Yoshihara Y, Watanabe I, Seto I (2010) A 77 GHz 90 nm CMOS transceiver for FMCW radar applications. IEEE J Solid-State Circuits 928–937
3. Sun J-T, He S-H, Liu Q, Liu H-W, Yoshimasu T (2011) Compact broadband Marchand balun with excellent imbalance performance for Si-based millimeter wave IC applications. In: China-Japan joint microwave conference proceedings (CJMW), 2011, pp 1–4, 20–22
4. Chiang M-J, Wu H-S, Tzuang C-K (2009) A compact CMOS Marchand balun incorporating meandered multilayer edge-coupled transmission lines. In: IEEE international microwave symposium, 2009, pp 125–128
5. Zhang Z, Guo Y-X, Ong LC, Chia MY (2004) Improved planar marchand balun using a patterned ground plane. Wiley Periodicals, Inc. pp 307–316
6. Liu J-X, Hsu C-Y, Chuang H-R, Chen C-Y (2007) A 60-GHz Millimeter-wave CMOS Marchand Balun. In: IEEE radio frequency integrated circuits (RFIC) symposium, 2007, pp 445–448
7. Lerdworatawee J, Won N (2005) Wide-band CMOS cascode low-noise amplifier design based on source degeneration topology. In: IEEE Trans Circuits and Systems I: Regular Papers, pp 2327–2334

Chapter 7
LNA Design for Variability

Nanoscale CMOS transistors are more susceptible to long-term electrical stress-induced reliability degradations. When those devices are used for radio frequency (RF) or microwave applications, a single transistor aging can lead to significant circuit performance degradation resulting from threshold voltage V_T shift and electron mobility μ_n drift. In addition, process variations in nanoscale transistors are another major concern in today's circuit design. Random dopant fluctuation, oxide thickness variation, and line edge roughness result in significant threshold voltage variation of CMOS transistors at sub-20 nm technology node and beyond [1].

The design for reliability (DFR) method intends to reduce the circuit over-design, while increasing its robustness against long-term aging. Here, the adaptive substrate (or body) bias scheme is described for the LNA design for process variability and circuit reliability [2]. Figure 7.1 shows a simple adaptive body bias scheme. The adaptive body bias technique dynamically adjusts the substrate bias of the input transistor $M1$ to reduce impact of process variations and device aging on circuit performance.

7.1 Analytical Model and Equations

As seen in Fig. 7.1, the right side of the circuit controls the substrate voltage of the main transistor. By designing similar drain-source voltage and gate-source voltage for $M1$ and $M2$, both the main transistor and bias transistor may subject to similar aging effect such as threshold voltage shift and electron mobility degradation.

© The Author(s) 2016
J.-S. Yuan, *CMOS RF Circuit Design for Reliability and Variability,*
SpringerBriefs in Reliability, DOI 10.1007/978-981-10-0884-9_7

Fig. 7.1 Adaptive body bias
design (© IEEE)

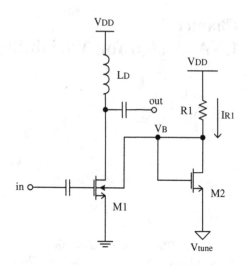

To account for possible different stress conditions between $M1$ and $M2$, mismatch
between the main transistor aging and bias transistor aging is also considered. In
Fig. 7.1, when the V_T of $M2$ increases, the current I_{R1} decreases. The reduced I_{R1}
results in an increased body voltage V_B. The increase in V_B of $M1$ will decrease the
threshold voltage of the input transistor due to source-body effect. Thus, this
compensates the change of V_T from device aging. Similarly, the decrease in electron
mobility, which decreases the drain current of the MOS transistor, will increase V_B
of $M1$. The drain current of $M1$ is also compensated. Examining Fig. 7.1, the KCL
equation to solve for V_B is given as

$$I_{R1}R1 + V_B = V_{DD} \tag{7.1}$$

$$I_{R1} \approx \frac{\beta'}{2}\left(V_B - V_{\text{tune}} - V_T'\right)^2 \tag{7.2}$$

where V_{tune} is the tuning voltage, β' is the transistor parameter ($\beta' = \mu_n C_{\text{ox}} W/L$) of
$M2$, and V_T' is the threshold voltage of $M2$. Note that V_{tune} can be used to adjust the
stress effect on $M2$ due to change of effective drain-source and gate-source voltages.
Combining (7.1) and (7.2) and solving for V_B one obtains

$$V_B = V_{\text{tune}} + V_T' + \frac{\sqrt{2\beta'R1(V_{DD} - V_{\text{tune}} - V_T') + 1} - 1}{\beta'R1}. \tag{7.3}$$

Using (7.3) the $\delta V_T'$ variation yields the body voltage fluctuation as follows:

$$
\begin{aligned}
\delta V_B &\approx \frac{\partial V_B}{\partial V_T'}\delta V_T' \\
&= (1 + \frac{-2\beta' R1}{2\beta' R1\sqrt{2\beta' R1(V_{DD} - V_{tune} - V_T') + 1}})\delta V_T' \\
&= \delta V_T' - \frac{\delta V_T'}{\sqrt{2\beta' R1(V_{DD} - V_{tune} - V_T') + 1}}
\end{aligned}
\tag{7.4}
$$

Due to the body effect, the V_T of $M1$ can be described by the following expression

$$
V_T = V_{T0} + \gamma_b(\sqrt{2\phi_F - V_B} - \sqrt{2\phi_F})
\tag{7.5}
$$

where γ_b is the body effect factor and ϕ_F represents the Fermi potential. The V_T shift of $M1$ due to degradation of $M1$ and $M2$ is thus modeled by the fluctuation of V_{T0} and V_B as

$$
\delta V_T = \delta V_{T0} - \frac{\gamma_b \delta V_B}{2\sqrt{2\phi_{FP} - V_B}}.
\tag{7.6}
$$

Combining (7.4) and (7.6) yields the V_T variation

$$
\delta V_T = \delta V_{T0} - \frac{\gamma_b \delta V_T'}{2\sqrt{2\phi_{FP} - V_B}}(1 - \frac{1}{\sqrt{2\beta' R1(V_{DD} - V_{tune} - V_T') + 1}}).
\tag{7.7}
$$

The first term in (7.7) represents the threshold voltage shift of $M1$, while the second term in (7.7) accomplishes the canceling effect resulting from the combination of threshold voltage shift of $M2$ and the body bias circuit of $M1$. Thus, the overall V_T shift of $M1$ due to process variation and reliability degradation is reduced. The level of reduction is related to $\delta V_T'$ of $M2$, body effect coefficient γ_b, $M2$ transistor β', and resistor $R1$. To achieve an optimal resilience to the variability and reliability, it is better to choose larger $R1$ and wider channel width of $M2$.

The noise factor is a measure of the degradation in signal-to-noise ratio that a system introduces. Equation (7.8) expresses the noise factor defined in the two-port network with noise sources and a noiseless circuit [3]. The noise figure is the noise factor expressed in decibels. The noise factor is written as

$$
F = \frac{\overline{i_s^2} + \overline{|i_n + (Y_c + Y_s)e_n|^2}}{\overline{i_s^2}} = 1 + \frac{\overline{i_n^2} + |Y_c + Y_s|^2\overline{e_n^2}}{\overline{i_s^2}}
\tag{7.8}
$$

Fig. 7.2 nMOSFET noise model

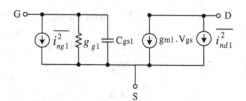

where i_s is the noise current from the source, Y_s is the source admittance, i_n is the device noise current, e_n is the device noise voltage, and Y_c is the correlation admittance.

For n-channel MOS transistor $M1$ at high frequency, the small-signal equivalent circuit model with noise currents is displayed in Fig. 7.2. The $1/f$ flicker noise is ignored at high frequency. The nMOSFET consists of the drain current noise and gate noise. The drain current noise and gate noise in Fig. 2 can be written as [4, 5]

$$\overline{i_{nd1}^2} = 4kT\gamma_1 g_{d01}\Delta f \tag{7.9}$$

$$\overline{i_{ng1}^2} = 4kT\theta\frac{\omega^2 C_{gs1}^2}{5g_{d01}}\Delta f \tag{7.10}$$

where k is the Boltzmanns' constant, T is the absolute temperature, ω is the radian frequency, g_{d01} is the output conductance of $M1$, C_{gs1} is the gate-source capacitance of $M1$, $\gamma_1 = 2/3$ for long channel MOSFET and can be 2–3 times larger in short-channel devices, and θ is the gate noise coefficient.

For the DFR biasing circuit, the drain of nMOSFET $M2$ is shorted to its gate as seen in Fig. 7.3. Thus, the noise looking into the node B consists of the two noise sources $R1$ and $M2$ drain current noise. The resistor $R1$ thermal noise and $M2$ drain current noise are modeled as:

Fig. 7.3 DFR biasing circuit noise model (© IEEE)

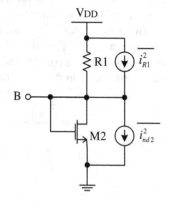

$$\overline{i_{R1}^2} = 4kT\frac{1}{R1}\Delta f \tag{7.11}$$

$$\overline{i_{nd2}^2} = 4kT\gamma_2 g_{d02}\Delta f \tag{7.12}$$

where g_{d02} is the output conductance of M2. Thus, the total mean squared noise voltage is

$$\overline{e_{B1}^2} = 4kT\frac{R1}{1+R1\gamma_2 g_{d02}}\Delta f. \tag{7.13}$$

The reflected drain current noise due to noise voltage in the body node is determined by a ratio of body transconductance g_{mb1}.

$$\overline{i_{nB1}^2} = 4kT\frac{R1}{1+R1\gamma_2 g_{d02}}g_{mb1}^2\Delta f. \tag{7.14}$$

Due to the body effect of M1, the drain current noise is a combination of noise originated from the drain current and reflected from the body node B.

$$\overline{i_{n1}^2} = \overline{i_{nB1}^2} + \overline{i_{nd1}^2} = 4kT\left[\frac{R1}{1+R1\gamma_2 g_{d02}}g_{mb1}^2 + \gamma_1 g_{d01}\right]\Delta f \tag{7.15}$$

The noise can be reflected back to the input gate of M1 by g_{m1}.

$$\overline{e_{n1}^2} = \frac{\overline{i_{n1}^2}}{g_{m1}^2} = 4kT\left[\frac{R1}{1+R1\gamma_2 g_{d02}}\frac{g_{mb1}^2}{g_{m1}^2} + \frac{\gamma_1 g_{d01}}{g_{m1}^2}\right]\Delta f \tag{7.16}$$

The equivalent input noise voltage is completely correlated with the drain current noise. Thus, the noise resistance is

$$R_{n1} = \frac{\overline{e_{n1}^2}}{4kT\Delta f} = \frac{R1}{1+R1\gamma_2 g_{d02}}\frac{g_{mb1}^2}{g_{m1}^2} + \frac{\gamma_1 g_{d01}}{g_{m1}^2} \tag{7.17}$$

The equivalent input noise voltage generator by itself does not fully account for the drain current noise. A noisy drain current also flows when the input is open circuited. Under this condition, the equivalent input voltage is obtained from dividing the drain current noise by the transconductance. When multiplying the input admittance, $\overline{e_{n1}^2}$ gives an equivalent input current noise as

$$\overline{i_{n1'}^2} = \overline{e_{n1}^2}(j\omega C_{gs1})^2 \tag{7.18}$$

Here, it is assumed that the input admittance of $M1$ is purely capacitive, which is good approximation when the operating frequency is below the cutoff frequency.

The drain noise and gate noise of $M1$ are correlated with a correlation coefficient c_1 defined as

$$c_1 = \frac{\overline{i_{ng1} \cdot i_{n1}^*}}{\sqrt{\overline{i_{ng1}^2} \cdot \overline{i_{n1}^2}}} \tag{7.19}$$

The total equivalent input current noise consists of the reflected drain noise and the induced gate current noise. The induced gate noise current itself has two parts. One part, i_{ngc1}, is fully correlated with the drain current noise of $M1$, while the other, i_{ngu1}, is uncorrelated with the drain current noise. The correlation admittance is expressed as follows:

$$Y_c = \frac{i_{n1'} + i_{ngc1}}{e_{n1}} = j\omega C_{gs1} + \frac{i_{ngc1}}{e_{n1}}$$

$$= j\omega C_{gs1} + g_{m1} \frac{i_{ngc1}}{i_{n1}} \tag{7.20}$$

The last term must be manipulated in terms of cross-correlations by multiplying both numerator and denominator by the conjugate of the drain current noise:

$$g_{m1} \frac{i_{ngc1}}{i_{n1}} = g_{m1} \frac{\overline{i_{ngc1} \cdot i_{n1}^*}}{\overline{i_{n1} \cdot i_{n1}^*}} = g_{m1} \frac{\overline{i_{ng1} \cdot i_{n1}^*}}{\overline{i_{n1}^2}} \tag{7.21}$$

Using the above equation, the correlation admittance can be rewritten as

$$Y_c = j\omega C_{gs1} + g_{m1} \frac{\overline{i_{ng1} \cdot i_{n1}^*}}{\overline{i_{n1}^2}}$$

$$= j\omega C_{gs1} + g_{m1} \frac{\overline{i_{ng1} \cdot i_{n1}^*}}{\sqrt{\overline{i_{ng1}^2}} \sqrt{\overline{i_{n1}^2}}} \sqrt{\frac{\overline{i_{ng1}^2}}{\overline{i_{n1}^2}}} = j\omega C_{gs1} + g_{m1} c_1 \sqrt{\frac{\overline{i_{ng1}^2}}{\overline{i_{n1}^2}}} \tag{7.22}$$

Inserting (7.10) and (7.15) into (7.22) yields the expression for Y_c. Note that the correlation coefficient c_1 is purely imaginary [3]. Thus, G_c (the real part of Y_c) equals zero. Using the definition of the correlation coefficient, the expression of the gate induced noise is written as

$$\overline{i_{ng1}^2} = \overline{(i_{ngc1} + i_{ngu1})^2} = 4kT\Delta f \left(\frac{\theta \omega^2 C_{gs1}^2 |c_1|^2}{5 g_{d01}} + \frac{\theta \omega^2 C_{gs1}^2 (1 - |c_1|^2)}{5 g_{d01}} \right). \tag{7.23}$$

Thus, the uncorrelated portion of the gate noise is

$$G_{ul} = \frac{\overline{i_{ul}^2}}{4kT\Delta f} = \frac{\theta\omega^2 C_{gs1}^2(1-|c_1|^2)}{5g_{d01}}.$$

(7.24)

The minimum noise figure is given by

$$F_{\min} = 1 + 2R_{n1}\left[G_{\text{opt}} + G_c\right] \approx 1 + 2R_{n1}\sqrt{\frac{G_{ul}}{R_{n1}}}$$

$$= 1 + \frac{2}{\sqrt{5}}\frac{\omega C_{gs1}}{g_{m1}}\sqrt{\theta(1-|c|^2)\left[\frac{R1g_{mb1}^2}{(1+R1\gamma_2 g_{d02})g_{d01}} + \gamma_1\right]}$$

(7.25)

Using (7.25) the minimum noise figure fluctuation is derived as

$$\Delta F_{\min} = -\frac{2}{\sqrt{5}}\frac{\omega C_{gs1}}{g_{m1}^2}\sqrt{\theta(1-|c|^2)\left[\frac{R1g_{mb1}^2}{(1+R1\gamma_2 g_{d02})g_{d01}} + \gamma_1\right]}\Delta g_{m1}$$

$$+ \frac{2}{\sqrt{5}}\frac{\omega C_{gs1}}{g_{m1}}\frac{\theta(1-|c|^2)R1g_{mb1}}{(1+R1\gamma_2 g_{d02})g_{d01}\sqrt{\theta(1-|c|^2)\left[\frac{R1g_{mb1}^2}{(1+R1\gamma_2 g_{d02})g_{d01}} + \gamma_1\right]}}\Delta g_{mb1}$$

$$- \frac{1}{\sqrt{5}}\frac{\omega C_{gs1}}{g_{m1}}\frac{\theta(1-|c|^2)R1g_{mb1}^2}{(1+R1\gamma_2 g_{d02})g_{d01}^2\sqrt{\theta(1-|c|^2)\left[\frac{R1g_{mb1}^2}{(1+R1\gamma_2 g_{d02})g_{d01}} + \gamma_1\right]}}\Delta g_{d01}$$

$$- \frac{1}{\sqrt{5}}\frac{\omega C_{gs1}}{g_{m1}}\frac{\theta(1-|c|^2)R1^2 g_{mb1}^2\gamma_2}{(1+R1\gamma_2 g_{d02})^2 g_{d01}\sqrt{\theta(1-|c|^2)\left[\frac{R1g_{mb1}^2}{(1+R1\gamma_2 g_{d02})g_{d01}} + \gamma_1\right]}}\Delta g_{d02}$$

(7.26)

In (7.26), the second term leads to the reduction of minimum noise figure sensitivity due to the body effect of MOSFET M1.

Small-signal gain S_{21} is related to the transconductance and gate-drain capacitance of $M1$. A detailed derivation of small-signal model is given in the following.

$$S_{21} = \frac{-2Y_{21}\sqrt{Z_{01}Z_{02}}}{\Delta_1}$$

(7.27)

$$\Delta_1 = (1 + Y_{11}Z_{01})(1 + Y_{22}Z_{02}) - Y_{21}Z_{01}Y_{12}Z_{02}$$

(7.28)

In the following discussion, one will see how Y_{21} fluctuates due to transconductance variation. Firstly, high frequency small-signal model for nMOSFET is shown in Fig. 7.4a. When the node D is tied to the ground terminal S, Fig. 7.4a reduces to Fig. 7.4b.

Fig. 7.4 **a** High frequency
small-signal model of
nMOSFET; **b** simplified
equivalent circuit for Y_{21}
derivation

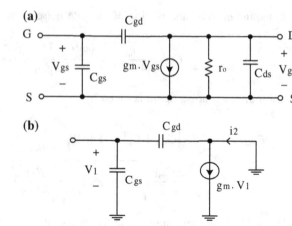

Y_{21} for single nMOSFET without body effect is derived from Fig. 7.4b. In Fig. 7.4, V_1 refers to V_{gs} in terminal 1 (between G and S) and V_2 refers to V_{gd} in terminal 2 (between D and S). Using Fig. 7.4b Y_{21} without body biasing is given by

$$Y_{21}(f) = \frac{i_2(f)}{V_1(f)}\big|_{V_2=0} = -j\omega C_{gd} + g_m \tag{7.29}$$

Thus, the transconductance fluctuation results in Y_{21} variation:

$$\Delta Y_{21}(f) = \Delta g_m \tag{7.30}$$

Figure 7.5a shows small-signal model for nMOSFET with body bias terminal. When D of $M1$ is tied to ground with S of both $M1$ and $M2$ in the substrate biasing circuit in Fig. 7.1, a simplified equivalent circuit model is displayed in Fig. 7.5b. Using Fig. 7.5b, one can write the current i_2

$$i_2 = g_m V_1 + g_{mb1} V_2 - V_1 j\omega C_{gd1}. \tag{7.31}$$

At the node B in Fig. 7.5b, the KCL equation results in

$$V_2 j\omega(C_{sb1} + C_{db1}) + V_2 j\omega(C_{gs2} + C_{ds2}) + g_{m2} V_2 + \frac{V_2}{R1\|r_{o2}} = (V_1 - V_2)j\omega C_{gb1} \tag{7.32}$$

Combining (7.31) and (7.32), Y_{21} is obtained:

$$Y_{21}(f) = \frac{i_2(f)}{V_1(f)}\big|_{V_2'=0} = -j\omega C_{gd1} + g_{m1} + \frac{j\omega C_{gb1} g_{mb1}}{j\omega C_{tot} + g_{m2} + \frac{1}{R1\|r_{o2}}} \tag{7.33}$$

where $C_{tot} = C_{sb1} + C_{db1} + C_{gs2} + C_{ds2}$.

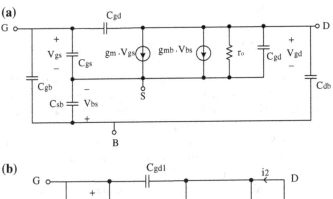

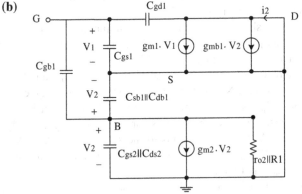

Fig. 7.5 **a** High frequency small-signal model of nMOSFET with body terminal and **b** small-signal model for Y_{21} derivation including substrate biasing circuit (© IEEE)

Note that V_2' in (7.33) represents what V_2 means in (7.29).

From (7.33) one can derive the fluctuation of Y_{21} as a function of g_{m1}, g_{mb1}, and g_{m2} as

$$\Delta Y_{21}(f) = \Delta g_{m1} - \frac{j\omega C_{gb1} g_{mb1}}{\left(j\omega C_{tot} + g_{m2} + \frac{1}{R1\|r_{o2}}\right)^2} \Delta g_{m2} + \frac{j\omega C_{gb1}}{j\omega C_{tot} + g_{m2} + \frac{1}{R1\|r_{o2}}} \Delta g_{mb1}$$

(7.34)

The second term in (7.34) will reduce Y_{21} sensitivity due to $M2$ in the DFR design. However, the third term in (7.34) due to the body effect of $M1$ will increase the fluctuation of Y_{21}. Thus, the transconductance of $M2$ helps reduce Y_{21} sensitivity, while the body transconductance of $M1$ may degrade Y_{21} sensitivity. Examining (7.26) and (7.34) together, the best sensitivity of noise figure and small-signal gain subject to body bias cannot be obtained simultaneously.

7.2 LNA Variability

A narrow-band cascode LNA designed at 24 GHz with adaptive body biasing is shown in Fig. 7.6. The main input transistor ($M1$) is connected with source degenerated inductor for better input matching and noise reduction. The cascode transistor ($M3$) provides the output to input isolation. All n-channel transistors are modeled using the PTM 65 nm technology [6]. The inductor values, MOS channel widths, and $R1$ are given in Fig. 7.6. $V_{DD} = 1.0$ V, $V_{bias} = 0.7$ V, and $R_{bias} = 5$ kΩ. The NF, NF_{min}, and S_{21} of the LNA without resilient biasing are 1.414, 1.226, and 12.124 dB at 24 GHz, while the corresponding values of the resilient design are 1.369, 1.327, and 11.531 dB, respectively.

Figures 7.7 and 7.8 show ADS Monte Carlo simulation [7] of the NF, NF_{min}, and S_{21} sensitivity subject to process variability. Monte Carlo simulation results demonstrate that a 10 % of V_T spread (STD/Mean) for the LNA without substrate biasing scheme yields 6.63 % NF spread and 5.58 % NF_{min} spread. A 10 % of V_T spread (STD/Mean) of the LNA with adaptive substrate biasing gives 3.85 % NF spread and 3.52 % NF_{min} spread. Comparing Figs. 7.7 and 7.8, it is apparent that the adaptive body biasing reduces the process variation effect significantly. It is also obtained that the ±0.2 V V_{tune} corresponds to the +5.41 to −4.16 % NF deviation and +5.20 to −3.92 % NF_{min} deviation. This spread fits into the compensation range for post-process V_{tune} calibration.

The reliability effect such as threshold voltage shift and mobility degradation on the LNA with or without adaptive substrate biasing is further evaluated. Figure 7.9 shows the normalized NF and NF_{min} to normalized threshold voltage shift for the

Fig. 7.6 A cascode low-noise amplifier with adaptive body bias (© IEEE)

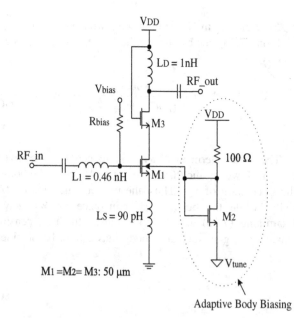

$M1 = M2 = M3$: 50 μm

Adaptive Body Biasing

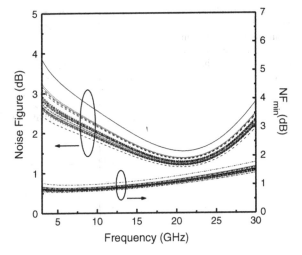

Fig. 7.7 Monte Carlo simulation of the LNA without substrate bias (© IEEE)

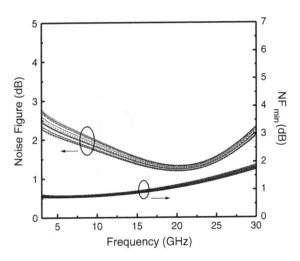

Fig. 7.8 Monte Carlo simulation of the LNA with the body bias technique (© IEEE)

original LNA compared to the LNA with adaptive bias design at different aging conditions. Since both drain-source voltage of main transistor $M1$ and substrate bias transistor $M2$ have the same designed drain-source voltage and similar gate-source voltage stress, $M1$ and $M2$ may have similar aging effect. However, different aging rates on $M1$ and $M2$ are also examined to account for a wide range of stress conditions. As seen in Fig. 7.9, the adaptive body biasing reduces the variation of normalized NF and NF_{min} significantly. In Fig. 7.9, the solid line represents the LNA without adaptive body bias and the solid lines with symbols represent the LNA with adaptive body bias, while the line with triangles corresponds to the $M2$ transistor's aging effect (threshold voltage shift or mobility degradation) is half of that of $M1$'s, the line with empty circles is when both $M1$ and $M2$ have an identical

Fig. 7.9 Normalized **a** NF and **b** NF$_{min}$ versus normalized V_T shift of the LNA without adaptive body bias compared to that with adaptive body bias (© IEEE)

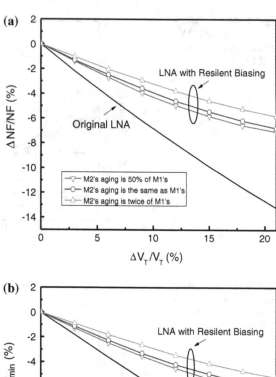

aging degradation, and the line with inverse triangles represents that $M2$'s aging effect is twice of $M1$'s. It is seen from Fig. 7.9 that the LNA with resilient substrate bias scheme reduces the noise figure and minimum noise figure sensitivity significantly even when the $M2$'s aging is different from that of $M1$'s. It is interesting to point out that larger $M2$ aging in fact reduces the noise figure sensitivity even further. This is due to an additional $\delta V_T'$ in $M2$ to compensate the threshold voltage shift δV_{T0} in $M1$ as indicated in Eq. (7.7).

Figure 7.10 shows the normalized NF and NF$_{min}$ variation versus normalized mobility degradation for the original LNA compared to the LNA with adaptive body bias at different mobility degradations. The line and symbol representations

Fig. 7.10 Normalized **a** NF and **b** NF_{min} versus normalized μ_n degradation of the LNA without adaptive bias compared to the LNA with adaptive body bias (© IEEE)

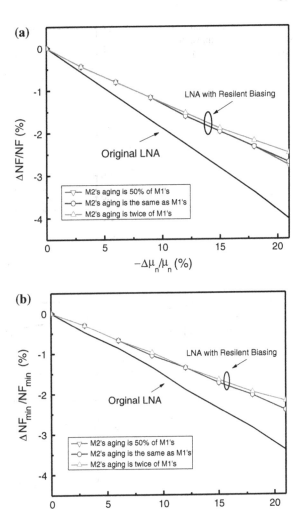

are the same as those defined in Fig. 7.9. The adaptive body biasing reduces the sensitivity of normalized NF and NF_{min} against mobility degradation also, though its effect is not as large as that in threshold voltage shift. With larger aging degradation on $M2$, the resilient biasing effect is further improved slightly.

The small-signal gain sensitivity versus V_T shift considering different aging is displayed in Fig. 7.11. Again, in this figure the solid line represents the LNA without adaptive body bias and the solid lines with symbols represent the LNA with adaptive body bias, while the triangles correspond to the $M2$ transistor's aging effect is half of that of $M1$'s, the empty circles are when both $M1$ and $M2$ have the

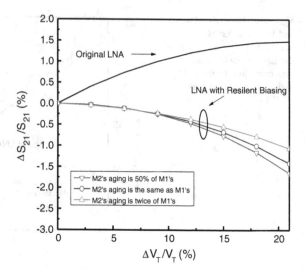

Fig. 7.11 Small-signal gain sensitivity versus threshold voltage shift (© IEEE)

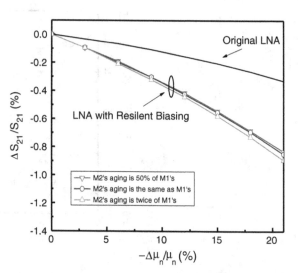

Fig. 7.12 Small-signal gain sensitivity versus electron mobility variation (© IEEE)

same aging degradation, and the inverse triangles represent $M2$'s aging effect twice of $M1$'s. In Fig. 7.11, the adaptive body biasing does not help reduce S_{21} sensitivity much as implied by Eq. (7.34). Figure 7.12 shows the normalized S_{21} sensitivity versus mobility degradation for the LNA with or without adaptive bias scheme. The adaptive body biasing increases the S_{21} sensitivity slightly subject to electron mobility degradation.

References

1. Chang L, Choi Y-K, Ha D, Ranade P, Xiong S, Bokor J, Hu C, King T-J (2003) Extremely scaled silicon nano-CMOS devices. Proc IEEE 91:1860–1873
2. Liu Y, Yuan JS (2011) CMOS RF low-noise amplifier design for variability and reliability. IEEE Trans Device Mater Reliab 11:450–457
3. Lee TH (1998) The design of CMOS radio-frequency integrated circuits. Cambridge University Press, Cambridge
4. van der Ziel A (1962) Thermal noise in field effect transistors. Proc IEEE 50:1801–12
5. van der Ziel A (1986) Noise in solid state devices and circuits. Wiley, New York
6. PTM web site. http://ptm.asu.edu/
7. http://www.agilent.com/find/eesof-ads

Chapter 8
Power Amplifier Design for Variability

It is clear that smaller feature size makes the MOSFET more sensitive to the process variations and stress-induced degradations. The circuit designer needs larger design margin to insure circuit robustness against such issues as yield and reliability. The process variability and reliability resilience design may reduce over design, while increase yield and circuit robustness. The resilient biasing technique aims to design reliable circuits capable of post-process adjustment and insensitive to the transistor parameter degradations over long-term stress effect.

Figure 8.1 shows a simplified variability and reliability resilient biasing design for the power amplifier, which introduces tunable adaptive body biasing.

The right branch of the circuit in Fig. 8.1 controls the body potential of the MOSFET $M1$. Thus, the threshold voltage of $M1$ can be adjusted by the body bias. The voltage source V_{tune} is used for post-fabrication calibration. During the long-term usage, both $M1$ and $M2$ are subject to similar reliability induced threshold voltage and electron mobility shifts. When the V_T of $M2$ increases, the branch current I_{R1} will decrease. The reduction in the branch current leads to an increase in the node voltage V_B. Therefore, the V_T of $M1$ will decrease due to combined reliability degradation and body effect. Similar mechanism applies to electron mobility degradation on both transistors. The drain current of $M1$ is thus more stable because of resilient biasing design scheme.

© The Author(s) 2016
J.-S. Yuan, *CMOS RF Circuit Design for Reliability and Variability*,
SpringerBriefs in Reliability, DOI 10.1007/978-981-10-0884-9_8

Fig. 8.1 Tunable adaptive body biasing (© IEEE)

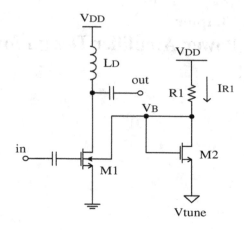

8.1 Analytical Model and Equations

Again, using the approach laid out in Chap. 7, the V_T shift of $M1$ due to degradations of both $M1$ and $M2$ is given by

$$\delta V_T = \delta V_{T0} - \frac{\gamma \cdot \delta V_T'}{2\sqrt{2\phi_{FP} - V_B}} \left(1 - \frac{1}{\sqrt{2\beta' R1(V_{DD} - V_{tune} - V_T') + 1}} \right). \quad (8.1)$$

The mobility degradation results in a decrease in drain current also. The drain current of $M1$ is simplified as $I_D \approx \beta(V_{GS} - V_T)^2/2$, where β variation due to mobility degradation is given by

$$\delta\beta = C_{ox}\frac{W}{L}\delta\mu_n. \quad (8.2)$$

Clearly, β variation is linearly proportion to the electron mobility drift. The same relationship also applies to β'. The node voltage V_B fluctuation due to mobility degradation is simplified to $\delta V_B \approx \frac{\partial V_B}{\partial \beta'}\delta\beta'$. Using (8.2) $\frac{\partial V_B}{\partial \beta'}$ is derived below:

$$\begin{aligned}
\frac{\partial V_B}{\partial \beta'} &= \frac{1}{(\beta' R1)^2} \left(\frac{2R1(V_{DD} - V_{tune} - V_T')\beta' R1}{2\sqrt{2\beta' R1(V_{DD} - V_{tune} - V_T') + 1}} - \sqrt{2\beta' R1(V_{DD} - V_{tune} - V_T') + 1} \cdot R1 \right) \\
&= \frac{\beta' R1^2(V_{DD} - V_{tune} - V_T') - 2\beta' R1^2(V_{DD} - V_{tune} - V_T') - R1'}{(\beta' R1)^2\sqrt{2\beta' R1(V_{DD} - V_{tune} - V_T') + 1}} \\
&= \frac{-R1(\beta' R1(V_{DD} - V_{tune} - V_T') + 1)}{(\beta' R1)^2\sqrt{2\beta' R1(V_{DD} - V_{tune} - V_T') + 1}}.
\end{aligned}$$

$$(8.3)$$

From the result in (8.3), one therefore finds δV_B as

$$\delta V_B = -\frac{R1(\beta'R1(V_{DD} - V_{tune} - V_T') + 1)}{(\beta'R1)^2 \sqrt{2\beta'R1(V_{DD} - V_{tune} - V_T') + 1}} \delta\beta'. \tag{8.4}$$

Assuming $\beta'R1(V_{DD} - V_{tune} - V_T') \gg 1$, (8.4) reduces to

$$\delta V_B \approx -\sqrt{\frac{V_{DD} - V_{tune} - V_T'}{2\beta'^3 R1}} \delta\beta'. \tag{8.5}$$

The threshold voltage variation in $M1$ due to body voltage fluctuation resulting from the mobility degradation in $M2$ is approximately as

$$\delta V_T \approx -\frac{\gamma \cdot \delta V_B}{2\sqrt{2\phi_{FP} - V_B}}. \tag{8.6}$$

The drain current fluctuation subject to key transistor parametric drifts ($\delta\beta$ and δV_T) is given by

$$\delta I_D = \frac{\partial I_D}{\partial \beta} \delta\beta + \frac{\partial I_D}{\partial V_T} \delta V_T. \tag{8.7}$$

In the derivation of $\frac{\partial I_D}{\partial \beta}$ and $\frac{\partial I_D}{\partial V_T}$, a simple drain current equation $\left(I_D \approx \frac{\beta}{2}(V_{GS} - V_T)^2\right)$ is used. The drain current variation is thus obtained as

$$\frac{\partial I_D}{\partial \beta} = \frac{1}{2}(V_{GS} - V_T)^2. \tag{8.8}$$

$$\frac{\partial I_D}{\partial V_T} = -\beta(V_{GS} - V_T) \tag{8.9}$$

Using (8.7), (8.8), and (8.9) one obtains

$$\delta I_D = \frac{1}{2}(V_{GS} - V_T)^2 \delta\beta - \beta(V_{GS} - V_T)\delta V_T. \tag{8.10}$$

Combining (8.2), (8.8), and (8.10), the fluctuation of drain current of $M1$ is expressed below

$$\delta I_D = \frac{1}{2}(V_{GS} - V_T)^2 \delta\beta - \beta(V_{GS} - V_T)\frac{\gamma\sqrt{\frac{V_{DD} - V_{tune} - V_T'}{2\beta'^3 R1}}\delta\beta'}{2\sqrt{2\phi_{FP} - V_B}}. \tag{8.11}$$

Note that the variation $\delta\beta$ reflects the fluctuation resulting from the electron mobility degradation of $M1$. $\delta\beta'$ represents the fluctuation caused by the electron mobility degradation of $M2$. The reduction of $M1$'s mobility will decrease the drain current in $M1$, while the reduction of $M2$'s mobility will increase the drain current in $M1$. To maximize the canceling effect, larger value of $R1$ as well as larger size of $M2$ are expected.

8.1.1 Tuning for Variability

The V_T shift of $M1$ due to V_{tune} change is described as follows. From (8.1) the body voltage values corresponding to the two different tuning voltages are determined by the equations in (8.12) and (8.13). Here, the V_T of $M2$ is supposed to be constant.

$$V_{B1} = V_{\text{tune1}} + V_T' + \frac{\sqrt{2\beta R1(V_{DD} - V_{\text{tune1}} - V_T') + 1} - 1}{\beta R1} \tag{8.12}$$

$$V_{B2} = V_{\text{tune2}} + V_T' + \frac{\sqrt{2\beta R1(V_{DD} - V_{\text{tune2}} - V_T') + 1} - 1}{\beta R1} \tag{8.13}$$

where V_{tune1} and V_{tune2} represent the two different tuning voltages.

The threshold voltage of $M1$ under the two different V_{tune} voltages can be written as:

$$V_{T1} = V_{T0} + \gamma_b\left(\sqrt{2\phi_F - V_{B1}} - \sqrt{2\phi_F}\right) \tag{8.14}$$

$$V_{T2} = V_{T0} + \gamma_b\left(\sqrt{2\phi_F - V_{B2}} - \sqrt{2\phi_F}\right). \tag{8.15}$$

The difference between two tuning voltage is marked as ΔV_T

$$\Delta V_T = V_{T2} - V_{T1}. \tag{8.16}$$

Combining (8.14) to (8.16), the sensitivity of V_T in $M1$ due to the tuning voltage of the circuit is derived as

$$\Delta V_T = \gamma_b\left(\sqrt{2\phi_F - V_{B2}} - \sqrt{2\phi_F - V_{B2}}\right). \tag{8.17}$$

A complete expression of (8.17) is complicated when substituting V_{B1} and V_{B2} with (8.12) and (8.13). Using (8.17) and the PTM 65 nm nMOSFET model parameters, the relationship between the threshold voltage and tuning voltage is calculated and plotted in Fig. 8.2.

Fig. 8.2 Normalized ΔV_T versus V_{tune} (© IEEE)

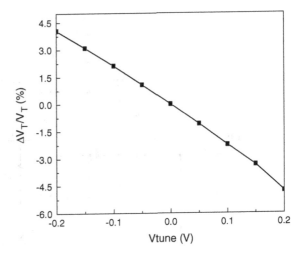

The V_T of $M1$ decreases linearly from 4.05 to −4.76 % as Vtune increases from −0.2 to 0.2 V. This property can serve as post-fabrication calibration to compensate for the V_T deviation of $M1$ due to process variability.

Both the fabrication process-induced fluctuation and time-dependent degradation cause the MOSFET model parameter shifts. V_T is the most significant parameter for the MOSFET suffering from variability and reliability degradations. Static post-fabrication calibration and dynamic V_T adjustment are considered using the resilient biasing design. Figure 8.3 shows a 24 GHz class-AB PA topology. The resilient biasing is circled in this plot. The output matching network is tuned using ADS load-pull instrument to obtain the optimum value. The 65 nm NMOS transistors are modeled by the PTM equivalent BSIM4 model card. The transistor sizes, capacitor and inductor values, and supply voltage are given in this figure.

8.1.2   ADS Monte Carlo Simulation

The simulated P_{sat}, and η_{add} of the PA without resilient biasing are 10.28 dBm, 10.96 dBm, and 34.25 %, while the corresponding values of the resilient design shown reach 10.90 dBm, 11.22 dBm, and 34.59 %, respectively. The matching network remains the same between the two PA schematics. Figure 8.4 shows 20 overlapping samples of the output power and power-added efficiency variations due to process fluctuation [1]. It is observed from the Monte Carlo simulations that a 10 % of V_T spread (STD/Mean) will lead to 1.83 % P_{sat} spread and 1.05 % η_{add} spread. It is also seen from the simulation that the ±0.2 V and ±0.25 V V_{tune}

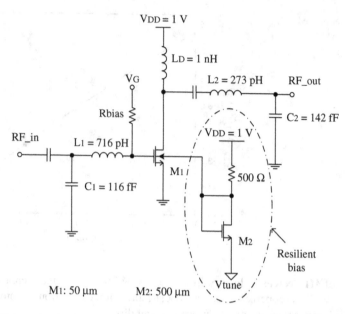

VDD = 1 V

LD = 1 nH

VG   L2 = 273 pH   RF_out

Rbias   C2 = 142 fF

RF_in   L1 = 716 pH   VDD = 1 V

C1 = 116 fF   M1   500 Ω

M2

Resilient
bias

M1: 50 μm   M2: 500 μm   Vtune

Fig. 8.3 Schematic of a 24 GHz class-AB power amplifier with resilient biasing (© IEEE)

correspond to the ±1.63 % and ±2.04 % P_{sat} deviation, respectively. So the spread fits into the compensation range of the ±0.25 V V_{tune} for post-process calibration.

The power amplifiers with and without resilient biasing technique are compared. Figure 8.5a shows normalized power-added efficiency to normalized threshold voltage variation. The resilient biasing reduces the sensitivity of normalized power-added efficiency significantly. For the normalized P_{sat} and P_{1dB} variations shown in Fig. 8.5b, the resilient biasing design reduces the sensitivity of P_{sat} and P_{1dB} against the threshold voltage shift dramatically, especially for the output power at the 1dB compression point (e.g., $\Delta P_{1dB}/P_{1dB}$ reduces from about −12 to −4 % at $\Delta V_T/V_T = 21$ %). So for reliability degradation induced dynamic V_T shift, the resilient biasing design helps improve the reliability of the PA by cutting the sensitivity by three to four times for the normalized output power at 1dB compression point and power-added efficiency.

The reliability degradation also reduces the electron mobility, which is another important parameter for drain current characteristic. Figure 8.6a shows normalized power-added efficiency versus normalized electron mobility reduction for PA with and without resilient biasing design. The resilient biasing scheme reduces the sensitivity of normalized power-added efficiency by 25 %. Figure 8.6b presents the normalized P_{sat} and P_{1dB} variations versus normalized mobility shift. The resilient design reduces the sensitivity of P_{sat} and P_{1dB} by 14.3 and 26.9 %, respectively. The resilient biasing design is obviously successful in reducing the power amplifier sensitivity against process variations and reliability degradations.

Fig. 8.4 PA performance
fluctuation of **a** output power
and **b** power-added efficiency
versus input power (© IEEE)

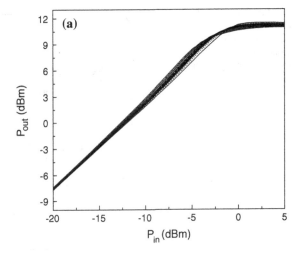

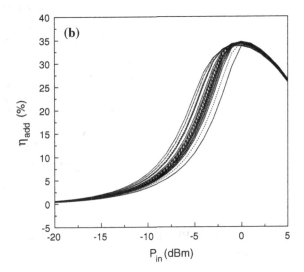

8.2   Use of Current Source for Sensing Variability

An on-chip variability sensor using current source [2] is studied to detect process,
supply voltage, and temperature (PVT) variations or even reliability degradation
stemming from hot electron effect. The PVT variations yield a control signal from
the designed current source. In Fig. 8.7, the current source circuit is made of
n-channel transistors $M1$, $M2$, and $M3$. The transistor $M1$ and $M2$ have the same
width and length and two times width of transistor $M3$. On the right branch in
Fig. 8.1, a resistor R is used to set a control voltage V_{Ctrl}. The reference current I_{ref}

Fig. 8.5 Normalized
a power-added efficiency
variation and **b** P_{sat} and P_{1dB}
variation versus normalized
threshold voltage shift
(© IEEE)

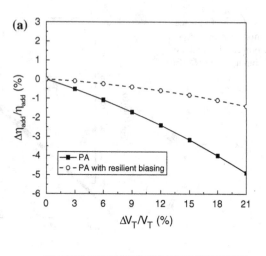

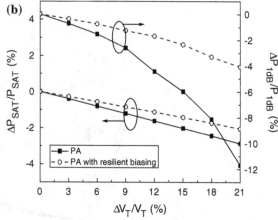

is dependent on the PVT fluctuations. The Kirchhoff's current law to solve for V_{Ctrl} is given by

$$V_{\text{Ctrl}} = V_{\text{DD}} - I_{\text{ref}}R \qquad (8.18)$$

and I_{ref} is the reference current and can be obtained as [3]

$$I_{\text{ref}} = \frac{(V_{\text{DD}} - V_{T1} - V_{T3})^2}{\left(\sqrt{\frac{2L_1}{K_n W_1}} + \sqrt{\frac{2L_3}{K_n W_3}}\right)^2} \qquad (8.19)$$

where K_n is the transconductance factor ($K_n = \mu_n \varepsilon_{ox}/t_{ox}$). Subscript 1 and 3 represent the transistor $M1$ and $M3$, respectively.

Fig. 8.6 Normalized **a** power-added efficiency variation and **b** P_{sat} and P_{1dB} variation versus normalized mobility shift (© IEEE)

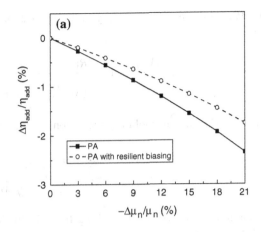

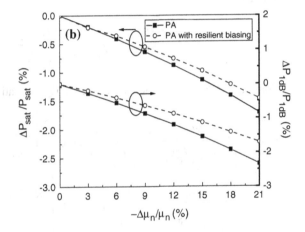

Fig. 8.7 Current source circuit schematic

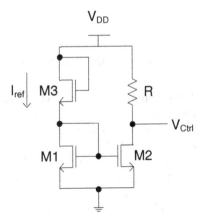

The V_{Ctrl} shift because of supply voltage variation is derived using (8.18) and (8.19)

$$\frac{\partial V_{\text{Ctrl}}}{\partial V_{\text{DD}}} = 1 - \frac{2R(V_{\text{DD}} - V_{T1} - V_{T3})}{\left(\sqrt{\frac{2L_1}{K_n W_1}} + \sqrt{\frac{2L_3}{K_n W_3}}\right)^2} \qquad (8.20)$$

The V_{Ctrl} shift due to mobility fluctuation is given by

$$\frac{\partial V_{\text{Ctrl}}}{\partial \mu_n} = -\frac{\varepsilon_{\text{ox}} R}{t_{\text{ox}}} \frac{(V_{\text{DD}} - V_{T1} - V_{T3})^2}{\left(\sqrt{\frac{2L_1}{W_1}} + \sqrt{\frac{2L_3}{W_3}}\right)^2} \qquad (8.21)$$

Furthermore, the V_{Ctrl} shift resulting from fluctuation of the threshold voltage from $M1$ or $M3$ is

$$\frac{\partial V_{\text{Ctrl}}}{\partial V_{T1,3}} = \frac{2R(V_{\text{DD}} - V_{T1} - V_{T3})}{\left(\sqrt{\frac{2L_1}{K_n W_1}} + \sqrt{\frac{2L_3}{K_n W_3}}\right)^2}. \qquad (8.22)$$

Combing (8.20)–(8.22) yields the overall V_{Ctrl} variation as follows:

$$\Delta V_{\text{Ctrl}} = \left[1 - \frac{2R(V_{\text{DD}} - V_{T1} - V_{T3})}{\left(\sqrt{\frac{2L_1}{K_n W_1}} + \sqrt{\frac{2L_3}{K_n W_3}}\right)^2} \right] \Delta V_{\text{DD}} - \left[\frac{\varepsilon_{\text{ox}} R}{t_{\text{ox}}} \frac{(V_{\text{DD}} - V_{T1} - V_{T3})^2}{\left(\sqrt{\frac{2L_1}{W_1}} + \sqrt{\frac{2L_3}{W_3}}\right)^2} \right] \Delta \mu_n$$

$$+ \left[\frac{2R(V_{\text{DD}} - V_{T1} - V_{T3})}{\left(\sqrt{\frac{2L_1}{K_n W_1}} + \sqrt{\frac{2L_3}{K_n W_3}}\right)^2} \right] \Delta V_{T1} + \left[\frac{2R(V_{\text{DD}} - V_{T1} - V_{T3})}{\left(\sqrt{\frac{2L_1}{K_n W_1}} + \sqrt{\frac{2L_3}{K_n W_3}}\right)^2} \right] \Delta V_{T3}$$

$$(8.23)$$

8.3 Tuning for Variability

The sensitivity of the class AB PA is evaluated using Fig. 8.8. The PVT variations change behaviors of the PA and also degrade the performance. In the simulation, the PVT variations are given to the PA circuit. Adaptive body biasing is used to find a range of body biasing voltage (V_{ABB}) to compensate each variation.

V_{Ctrl} signal is efficiently transformed to an optimal body bias signal for power amplifier application. From a range of V_{ABB}, an operational amplifier is used as a voltage shifter and amplifier to adjust the V_{Ctrl} to meet a required V_{ABB}. Choosing appropriate size of resistor R_1 and R_2 using (8.31) provides a matched V_{ABB} for PA.

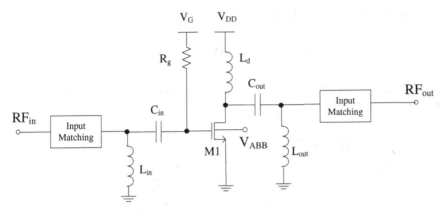

Fig. 8.8 A class AB PA with adaptive body biasing (© Elsevier)

Fig. 8.9 Level shifting
circuit

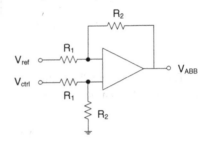

For example, for a reference voltage (V_{ref}) of 0.4 V, R_1 and R_2 can be designed at
500 Ω and 1500 Ω, respectively. (See Fig. 8.9)

$$V_{ABB} = \frac{R_2}{R_1}(V_{Ctrl} - V_{ref}) \tag{8.24}$$

Due to the body effect, the threshold voltage of the power amplifier transistor is
described by the following expression

$$V_T = V_{T0} + \gamma_b(\sqrt{2\phi_F - V_{ABB}} - \sqrt{2\phi_F}) \tag{8.25}$$

The threshold voltage shift of the PA transistor is modeled by the fluctuation of
V_{T0} and V_{ABB} as

$$\Delta V_T = \frac{\partial V_T}{\partial V_{T0}}\Delta V_{To} + \frac{\partial V_T}{\partial V_{ABB}}\Delta V_{ABB} = \Delta V_{T0} - \frac{\gamma}{2\sqrt{2\phi_F - V_{ABB}}}\Delta V_{ABB} \tag{8.26}$$

From (8.24), the V_{ABB} shift is given by

$$\Delta V_{ABB} = \frac{\partial V_{ABB}}{\partial V_{Ctrl}} \Delta V_{Ctrl} = \frac{R_2}{R_1} \Delta V_{Ctrl} \tag{8.27}$$

Thus, the threshold voltage shift of the power amplifier input transistor due to PVT variations are summed as

$$\Delta V_T = \Delta V_{T0} - \frac{\gamma_b R_2}{2R_1 \sqrt{2\phi_F - V_{ABB}}}$$

$$\left\{ \left[1 - \frac{2R(V_{DD} - V_{T1} - V_{T3})}{\left(\sqrt{\frac{2L_1}{K_n W_1}} + \sqrt{\frac{2L_3}{K_n W_3}} \right)^2} \right] \Delta V_{DD} - \left[\frac{\varepsilon_{ox} R}{t_{ox}} \frac{(V_{DD} - V_{T1} - V_{T3})^2}{\left(\sqrt{\frac{2L_1}{W_1}} + \sqrt{\frac{2L_3}{W_3}} \right)^2} \right] \Delta \mu_n \right.$$

$$\left. + \left[\frac{2R(V_{DD} - V_{T1} - V_{T3})}{\left(\sqrt{\frac{2L_1}{K_n W_1}} + \sqrt{\frac{2L_3}{K_n W_3}} \right)^2} \right] \Delta V_{T1} + \left[\frac{2R(V_{DD} - V_{T1} - V_{T3})}{\left(\sqrt{\frac{2L_1}{K_n W_1}} + \sqrt{\frac{2L_3}{K_n W_3}} \right)^2} \right] \Delta V_{T3} \right\}$$

$$\tag{8.28}$$

The drain current fluctuation subjects to key transistor parametric drifts $\Delta \mu_n$, ΔV_{GS} and ΔV_T can be modeled as

$$\Delta I_D = \frac{\partial I_D}{\partial \mu_n} \Delta \mu_n + \frac{\partial I_D}{\partial V_{GS}} \Delta V_{GS} + \frac{\partial I_D}{\partial V_T} \Delta V_T \tag{8.29}$$

Assume the V_{GS} shift is proportional to the fluctuation of V_{DD}.

$$\Delta V_{GS} = \alpha \Delta V_{DD} \tag{8.30}$$

where α is a fitting parameter.

Using (8.26)–(8.30) the fluctuation of drain current normalized to its fresh current is expressed as follows:

$$\frac{\Delta I_D}{I_D} = \frac{\Delta \mu_n}{\mu_n} + \frac{2\alpha_{DD}}{V_{GS} - V_T} - \frac{2}{V_{GS} - V_T} \left(\Delta V_{T0} - \frac{\gamma_b R_2}{2R_1 \sqrt{2\varphi_F - V_{ABB}}} \right.$$

$$\left\{ \left[1 - \frac{2R(V_{DD} - V_{T1} - V_{T3})}{\left(\sqrt{\frac{2L_1}{K_n W_1}} + \sqrt{\frac{2L_3}{K_n W_3}} \right)^2} \right]_{DD} - \left[\frac{\varepsilon_{ox}}{t_{ox}} \frac{2R(V_{DD} - V_{T1} - V_{T3})}{\left(\sqrt{\frac{2L_1}{W_1}} + \sqrt{\frac{2L_3}{W_3}} \right)^2} \right] \Delta \mu_n \right.$$

$$\left. \left. + \left[\frac{2R(V_{DD} - V_{T1} - V_{T3})}{\left(\sqrt{\frac{2L_1}{K_n W_1}} + \sqrt{\frac{2L_3}{K_n W_3}} \right)^2} \right] \Delta V_{T1} + \left[\frac{2R(V_{DD} - V_{T1} - V_{T3})}{\left(\sqrt{\frac{2L_1}{K_n W_1}} + \sqrt{\frac{2L_3}{K_n W_3}} \right)^2} \right]_{T3} \right\} \right)$$

$$\tag{8.31}$$

In the above equation, the terms beyond ΔV_{T0} represent the V_{DD}, mobility, and threshold voltage compensation effects. The normalized output power degradation is related to the normalized drain current degradation as follows [4]:

$$\frac{\Delta P_o}{P_o} \approx \frac{\Delta I_D}{I_D} \tag{8.32}$$

8.3.1 Circuit Simulation Results

The power amplifier with the current source compensation technique is compared with the PA without compensation using ADS simulation. For the process variation effect, the output power is evaluated against threshold voltage and mobility variations as shown in Figs. 8.10 and 8.11. It is clear from Figs. 8.10 and 8.11 that the power amplifier with adaptive body bias is more robust against threshold voltage variation (see Fig. 8.10) and mobility fluctuation (Fig. 8.11).

For the process variation effect, the output power of the PA has also been evaluated using different process corner models due to inter-die variations. The simulation result of the fast–fast, slow–slow, and nominal–nominal models is shown in Fig. 8.12. Clearly, the PA using the adaptive body bias compensation exhibits better stability against process variation effect.

Figures 8.13 and 8.14 show the output power of the power amplifier versus temperature variation and supply voltage change, respectively. As seen in Figs. 8.13 and 8.14 the output power of the PA using the adaptive body bias compensation technique demonstrates less sensitivity over temperature and V_{DD} variations.

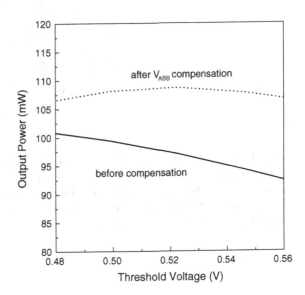

Fig. 8.10 Output power versus threshold voltage shift (© Elsevier)

Fig. 8.11 Output power
versus mobility variation
(© Elsevier)

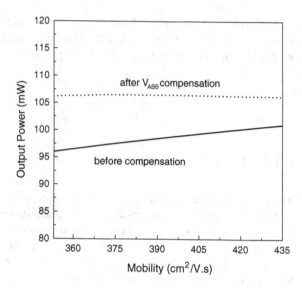

Fig. 8.12 Output power
versus process corner models
(© Elsevier)

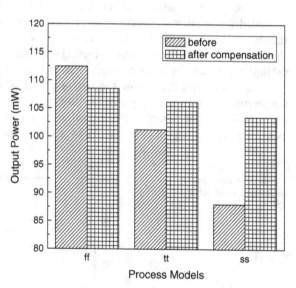

In addition, the power-added efficiency of the power amplifier with or without
adaptive body bias compensation is examined against semiconductor process
variations effects. Figures 8.15 and 8.16 display the improvement of power-added
efficiency of the PA with ABB compensation over that without adaptive body bias
for the threshold voltage shift (see Fig. 8.15) and mobility variation (see Fig. 8.16).

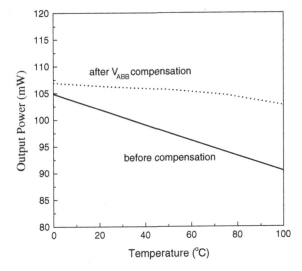

Fig. 8.13 Output power versus temperature (© Elsevier)

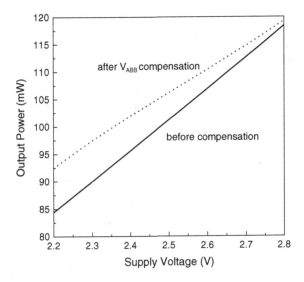

Fig. 8.14 Output power versus supply voltage (© Elsevier)

For the process corner models the power-added efficiency of the PA with ABB compensation shows less process sensitivity, as evidenced by the plot in Fig. 8.17.

Then, the power-added efficiency is compared against temperature and supply voltage variations. The power-added efficiency is getting better for the PA with ABB compensation as shown in Figs. 8.18 and 8.19.

Fig. 8.15 Power-added
efficiency as a function of
threshold voltage (© Elsevier)

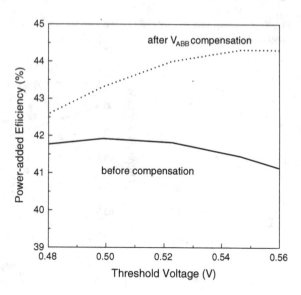

Fig. 8.16 Power-added
efficiency as a function of
mobility (© Elsevier)

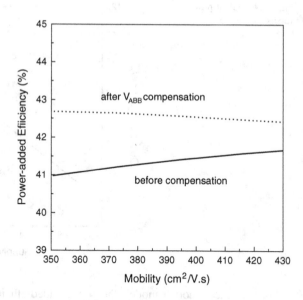

Fig. 8.17 Power-added
efficiency versus process
corner models (© Elsevier)

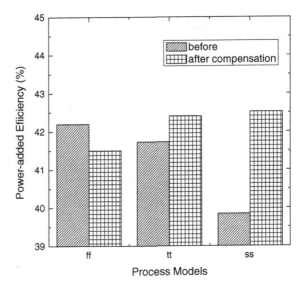

Fig. 8.18 Power-added
efficiency versus temperature
(© Elsevier)

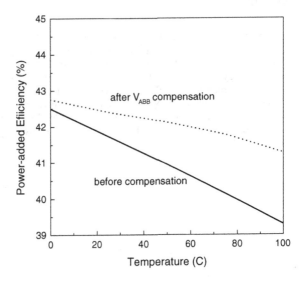

Fig. 8.19 Power-added
efficiency versus supply
voltage (© Elsevier)

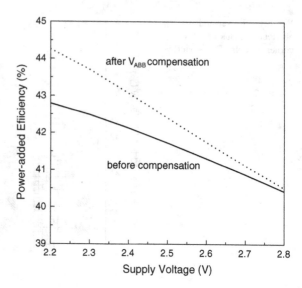

References

1. Liu Y, Yuan JS (2011) CMOS RF power amplifier variability and reliability resilience biasing
   design and analysis. IEEE Trans Electron Devices 540–546
2. Pappu AM, Zhang X, Harrison AV, Apsel AB (2007) Process invariant current source design:
   Methodology and examples. IEEE J Solid-State Circuits 2293–2302
3. Yuan JS, Kritchanchai E (2013) Power amplifier resilient design for process, voltage, and
   temperature variations. In: Microelectronics reliability, pp 856–860
4. Quemerais T, Moquillon L, Huard V, Fournier J-M, Benech P, Corrao N, Mescot X (2010)
   Hot-carrier stress effect on a CMOS 65-nm 60-GHz one-stage power amplifier. IEEE Electron
   Device Lett 927–929

Chapter 9
Oscillator Design for Variability

9.1 Mixed-Mode Device and Circuit Simulation

To evaluate the physical insight into the Colpitts oscillator circuit operation, the mixed-mode device and circuit simulation using Sentaurus TCAD software [1] is adopted. In Sentaurus device simulation, Poisson's and continuity equations with drift-diffusion transport are implemented. The Shockley–Read–Hall carrier recombination, Auger recombination, and impact ionization models are used. The physical model for impact ionization used in this work is the University of Bologna impact ionization model, based on impact ionization data generated by the Boltzmann solver [2]. It covers a wide range of electric fields (50–600 kV/cm) and temperatures (300–700 K). It is calibrated against impact ionization measurements in the whole temperature range [3]. The low field mobility is calculated by Mathiessen's rule and incorporates the bulk and surface mobility. To account for lattice heating, Thermodynamic, Thermode, RecGenHeat, and AnalyticTEP models in Sentaurus are included. The thermodynamic model extends the drift-diffusion approach to account for electrothermal effects. A Thermode is a boundary where the Dirichlet boundary condition is set for the lattice. RecGenHeat includes generation–recombination heat sources. AnalyticTEP gives the analytical expression for thermoelectric power.

Figure 9.1 shows the Colpitts oscillator used in the mixed-mode device and circuit simulation [4]. The mixed-mode simulation provides the device physical insight and response in the practical circuit environment. In Sentaurus simulation, the MOSFET has the channel length of 65 nm and the channel width of 64 μm. The circuit parameters used are C_1 = 22 pF, C_2 = 27.2 pF, L_D = 0.15 nH, R_D = 2900 Ω, R_S = 40 Ω, V_G = 1.8 V, and V_{DD} = 3.3 V. The simulated oscillator output response from Sentaurus is displayed in Fig. 9.2. The oscillator has a sinusoidal oscillating

© The Author(s) 2016
J.-S. Yuan, *CMOS RF Circuit Design for Reliability and Variability*,
SpringerBriefs in Reliability, DOI 10.1007/978-981-10-0884-9_9

Fig. 9.1 Schematics of an oscillator used in the mixed-mode device and circuit simulation

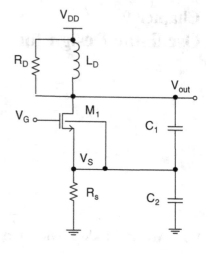

Fig. 9.2 Oscillator output response from mixed-mode device and circuit simulation (© IEEE)

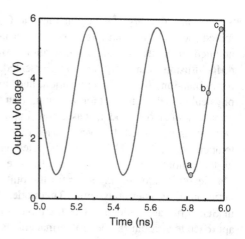

output waveform from 0.5 to 2.9 V. To analyze the reliability effect on the Colpitts oscillator, gate-source voltage and drain-source voltage as a function of time are depicted in Fig. 9.3. Examining the voltage waveforms in Fig. 9.3, one can define three key points a, b, and c (i.e., the bottom, middle, and top of the output voltage) to probe impact ionization and self-heating at these three critical time points. The I.I. rates, electric field, and total current density from Sentaurus mixed-mode device and circuit simulation are plotted in Figs. 9.4, 9.5, and 9.6, respectively. Note that no lattice heating of the Colpitts oscillator was observed in mixed-mode device and circuit simulation (data not shown).

To investigate the physical insight into hot electron injection, impact ionization rates at the three different time points are shown in Fig. 9.4. At point a the drain-source voltage reaches the minimum and its corresponding electric field is low; however, the current density is very high due to large V_{GS} (see Fig. 9.5). At

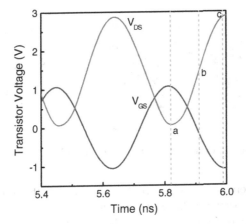

Fig. 9.3 Gate-source and drain-source voltages versus time (© IEEE)

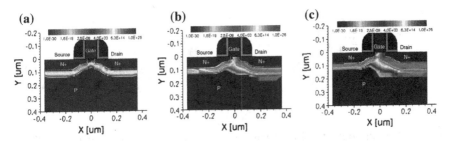

Fig. 9.4 Impact ionization rates at points *a*, *b*, and *c* in Fig. 9.2 (© IEEE)

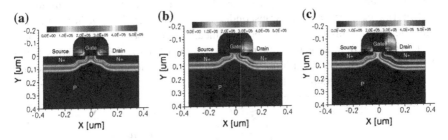

Fig. 9.5 Electric field at points *a*, *b*, and *c* in Fig. 9.2 (© IEEE)

point *c*, the I.I. rates are high because the drain-source voltage reaches the maximum. The impact ionization rates at point *b* are higher than those at points *a*. This is attributed to relatively high drain-source voltage and drain current density at point *b*, as indicated in Fig. 9.3. Higher drain current enhances I.I. generated carriers under high electric field. The peak impact ionization rates at points *b* and *c* reach 1×10^{26}/cm^3/s, a precursor of hot carrier effect. The hot electron reliability issue

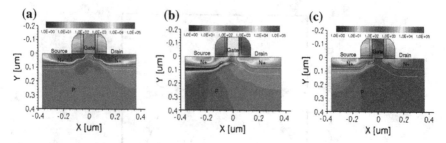

Fig. 9.6 Total current density at points *a*, *b*, and *c* in Fig. 9.2 (© IEEE)

becomes even more important when the channel length of the nMOSFET is decreasing and the supply voltage of the circuit is increasing.

The phase noise of Colpitts oscillator shown in Fig. 9.1 is analyzed for examining device parameter variations. The phase noise to account for MOS transistor parameter drift due to aging is expressed as [5]:

$$L(\Delta f) = 10 \log \left(\frac{\bar{V}_n^2}{2 V_{\text{tank}}^2} \right)$$
$$= 10 \log \left\{ \left[\left(\frac{|g_{m(1)}|^2 K_f}{4 C_{\text{ox}} W L \Delta f} + \sum_{n=1}^{\infty} |g_{m(n-1)} + g_{m(n+1)}|^2 \times \frac{kT\gamma}{\bar{g}_m} \right) \alpha + \frac{kT}{R} \right] \left(\frac{R f_0}{Q \Delta f A_S} \right)^2 \right\}$$

(9.1)

where V_n is the output noise voltage, V_{tank} is the signal voltage of the oscillator output, $g_{m(n)}$ is the *n*th Fourier coefficients of transconductance, K_f is a process dependent constant on the order of 10^{-25} V^2F, f_0 is the center frequency, γ is a coefficient (about 2/3 for long-channel transistors and larger for submicron MOSFETs), $\bar{g}_m$ is the average transconductance of the transistor, α is the transfer parameter from nonlinear network port to linear network port ($\alpha = (1 - F)^2$, where $F = C_1/(C_1 + C_2)$), R is the parasitic resistance in the LC tank, Q is the quality factor of LC tank, and A_S is the amplitude of the AC voltage at the source of the transistor.

In (9.1) $\bar{g}_m = \beta A_S (\sin \theta - \theta \cos \theta)/\pi$, $\beta = \mu_{n0} C_{\text{ox}} W/L$, $\theta = \cos^{-1}[(V_T - V_G)/A_S]$, $\beta = \mu_{n0} C_{\text{ox}} W/L$, $g_{m(n)} = g_{m(-n)}$, and

$$g_{m(n)} = \begin{cases} \beta A_S \left[\frac{(\sin \theta - \theta \cos \theta)}{\pi} \right] & \text{for } n = 0 \\ \beta A_S \left[\frac{(\theta - \sin \theta \cos \theta)}{\pi} \right] & \text{for } n = 1 \\ 2 \beta A_S \left[\frac{\sin n\theta \cos \theta - n \cos \theta \sin \theta}{n(n^2-1)\pi} \right] & \text{for } n \geq 2 \end{cases}$$

(9.2)

The Colpitts oscillator shown in Fig. 9.1 has been simulated in ADS. To be consistent with the mixed-mode simulation condition, the same circuit element values used in mixed-mode simulation are also used in the ADS circuit simulation. The simulated output waveform as a function of time and its Spectral density versus

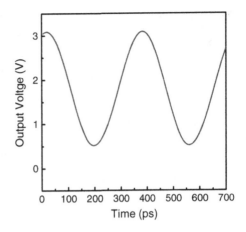

Fig. 9.7 Simulated output waveform versus time

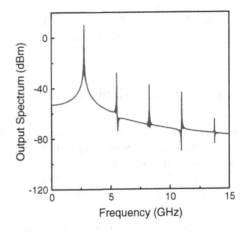

Fig. 9.8 Simulated output power spectrum characteristics

frequency are depicted in Figs. 9.7 and 9.8. The oscillation frequency measured from Fig. 9.7 is 2.4 GHz and its fundamental signal Spectral power is −4 dBm at 2.4 GHz. The phase noise predicted by the analytical equation in (9.1) is compared with that by the ADS simulation result in Fig. 9.9. In Fig. 9.9, the solid circles represent the model predictions and the solid line represents the ADS simulation. A good agreement between the model predictions and ADS simulation results before hot electron stress is obtained. The HCI effect on the phase noise is also displayed using K_f factor in (9.1). As seen in Fig. 9.9 the phase increases with increasing K_f factor, which is related to the interface quality or interface states between the SiO_2 and Si interface. Intuitively, the worse the process condition, the larger is the interface states. The longer the stress time, the larger is the interface trap density [6]. If the analytical equation on process and stress dependent K_f factor becomes available, Eq. (9.1) can account for more process and stress effects

Fig. 9.9 Phase noise modeling versus offset frequency including K_f effect

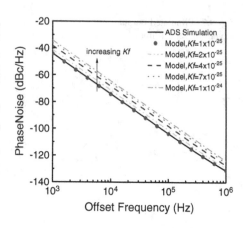

inclusively. As hot carriers generate more interface states at the SiO_2 and Si interfacial layer, the K_f factor increases, thus the $1/f$ noise of the MOSFET and phase noise of the oscillator increase.

9.2 Process Variability and Adaptive Body Bias

RDF [7] remains the dominant source of statistical variability and is mainly caused by silicon dopant fluctuations during fabrication process. It becomes more severe as device size shrinks. LER [8], a random deviation of line edges from gate definition, does not scale with line width. PGG [9] is attributed to gate dielectric thickness variations which contribute to threshold voltage variation. All the above mentioned process variations cause fluctuation of threshold voltage, mobility, and oxide thickness, which in turn affect the device and circuit performance. Furthermore, reliability issue could widen the standard derivation of process variation in Gaussian distribution [10].

To further examine the process variation and reliability impact on Colpitts oscillator, Monte Carlo (MC) circuit simulation has been performed. In ADS, the Monte Carlo simulation assumes statistical variations (Gaussian distribution) of transistor model parameters such as the threshold voltage, mobility, and oxide thickness. After Monte Carlo simulations with a sample size of 1000, the phase noise variation is displayed in Fig. 9.10. In this histogram plot, the x-axis shows the phase noise distribution of the oscillator without body effect and y-axis displays the probability density of occurrence. The mean value of phase noise is -121 dBc/Hz and the standard deviation is 0.71 dBc/Hz. In Monte Carlo simulation, the initial values of V_{T0}, μ_0, and t_{ox} are 0.42 V, 491 cm^2/V s, and 1.85 nm, respectively. The statistical variations for V_{T0}, μ_0, and t_{ox} are set at ± 10, ± 5, and ± 3 % from 65 nm technology node.

Fig. 9.10 Phase noise
distribution @ $\Delta f = 400$ kHz
(© IEEE)

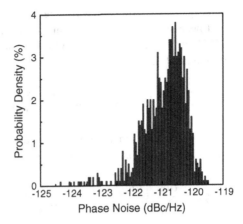

To reduce the process variation effect on the Colpitts oscillator, an adaptive body bias scheme as shown in Fig. 9.11 is proposed. In Fig. 9.11, the body bias of $M1$ is determined by the adaptive body bias circuit in the dashed oval circle.

To account for the body bias effect, the threshold voltage of $M1$ can be written as

$$V_T = V_{T0} + \gamma_b \left(\sqrt{2\phi_F - V_{BS}} - \sqrt{2\phi_F} \right) \tag{9.3}$$

Fig. 9.11 Colpitts oscillator
with adaptive body bias
(© IEEE)

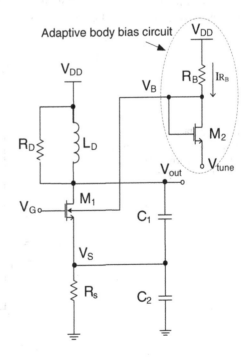

where V_{BS} is the body-source voltage. Note that the source voltage V_S in this circuit is not necessarily equal to zero, unlike the case of power amplifiers in Chap. 8.

The drain current of the MOSFET including the body bias effect can be approximated as

$$I_{DS} \approx \frac{\mu_{n0} C_{ox} W}{2L[1 + \theta_1(V_{GS} - V_T) + \theta_2 V_{BS}]} (V_{GS} - V_T)^2 \tag{9.4}$$

The transconductance based on the derivative of drain current with respect to gate-source voltage is derived as

$$g_m \equiv \frac{\partial I_{DS}}{\partial V_{GS}} = \frac{\mu_{n0} C_{ox} W}{2L} \frac{2(V_{GS} - V_T)(1 + \theta_2 V_{BS}) + \theta_1(V_{GS} - V_T)^2}{[1 + \theta_1(V_{GS} - V_T) + \theta_2 V_{BS}]^2} \tag{9.5}$$

The transconductance equation taking the body bias into account in (9.5) is then used in the phase noise prediction in (9.1) for the oscillator with an adaptive body bias.

The MC simulation result of the Colpitts oscillator including the body bias effect is shown in Fig. 9.12. In this histogram plot, the mean value of phase noise is -121 dBc/Hz and the standard deviation is 0.18 dBc/Hz. The phase noise is evaluated at the offset frequency of 400 kHz.

Comparing Figs. 9.10 and 9.12, the adaptive body bias clearly reduces the oscillator process sensitivity significantly. The use of adaptive body bias to reduce process variability effect on the Colpitts oscillator can be explained as follows: The threshold voltage shift including body bias effect can be expressed as

$$\Delta V_T = \Delta V_{T0} - \frac{\gamma_b \times \Delta V_{BS}}{2\sqrt{2\phi_F - V_{BS}}} \tag{9.6}$$

Fig. 9.12 Phase noise distribution using the adaptive body bias scheme (Δf is at 400 kHz) (© IEEE)

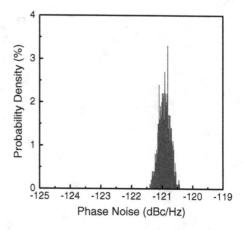

where ΔV_{T0} is the threshold voltage change resulting from process variations and ΔV_{BS} is produced due to current variation and the use of body bias circuit. The minus sign of the second term in (9.6) indicates that the body bias effect provides a compensation effect for threshold voltage variations from process uncertainties. Again, when the process variations increase the threshold voltage of $M1$ in Fig. 9.11, the body bias V_B to $M1$ increases due to less $I_{D2}R_B$ ohmic loss in the adaptive body bias circuit. This tends to decrease the V_T in $M1$ to compensate the initial increase in V_T. On the other hand, when process variability decreases the V_T in $M1$, the adaptive body bias circuit will decrease the V_B to $M1$. This in turn increases the V_T in $M1$ to compensate the initial decrease in V_T.

References

1. http://www.synopsys.com
2. Vecchi MC, Rudan M (1998) Modeling electron and hole transport with full-band structure effects by means of the spherical-harmonics expansion of the BTE. IEEE Trans. Electron Devices 230–238
3. Valdinoci et al. (1999) Impact-ionization in silicon at large operating temperature. In: International conference simulation of semiconductor processes and devices (SISPAD), Kyoto, Japan, 1999, pp 27–30
4. Yuan JS, Chen S (2012) A simulation study of Colpitts oscillator reliability and variability. IEEE Trans. Device Mater Reliab 576–581
5. Li Z, Bao J, Tang P, Wang F (2010) Analysis of phase noise spectrum in LC oscillator based on nonlinear method. In: International conference on microwave and millimeter wave technology, 2010, pp 649–652
6. Chen Z, Garg P, Ong AP (2003) Reduction of hot-carrier-induced 1/f noise of MOS devices using deuterium processing. In: IEEE conference electron devices and solid-state circuits, 2003, pp 197–199
7. Asenov, Simulation of statistical variability in Nano MOSFETs. In: Symposium of VLSI Technology, 2007, pp 86–87
8. Sugii N, Tsuchiya R, Ishigaki T, Morita Y, Yoshimoto H, Kimura S (2010) Local V_{th} variability and scalability in silicon-on-thin-box (SOTB) CMOS with small random-dopant fluctuation. IEEE Trans Electron Devices 835–845
9. Li Y, Hwang C-H, Li T-Y (2009) Random-dopant-induced variability in nano-CMOS devices and digital circuits. IEEE Trans Electron Devices 1581–1597
10. Gómez D, Sroka M, Luis J, Jiménez G (2010) Process and temperature compensation for RF low-noise amplifiers and mixers. In: IEEE Trans Circuits and Systems-I: Regular Papers, pp 1204–1211

Chapter 10
Mixer Design for Variability

Both the fabrication process-induced fluctuation and time-dependent degradation cause the MOSFET model parameters to drift. The threshold voltage and mobility are the two most significant model parameters that suffer from process uncertainty and reliability degradations. Here, the most widely used double-balanced Gilbert structure [1] in Fig. 10.1 is used to evaluate the process variations and aging effects on RF mixer performance. In this figure, positive and negative RF input signals are applied to transistors $M1$ and $M2$. Local oscillator (LO) signals are applied to switching transistors $M3$, $M4$, $M5$, and $M6$. The transistor $M7$ provides the bias current. RF and LO multiplication produces the output signal at intermediate frequency (IF).

The conversion gain (CG) of the mixer can be derived as

$$CG = \frac{2}{\pi} \frac{R_L}{R_S + \frac{1}{g_m}} \tag{10.1}$$

where R_L is the load resistance and R_S is the inductor resistance. The noise figure (NF) of the mixer is given by

$$NF = 10 \log_{10}(F) \tag{10.2}$$

where F is the flicker noise, which is derived as

$$F = \frac{\pi^2}{4} \left(1 + \frac{2\gamma_1}{g_m R_S} + \frac{2}{g_m^2 R_L R_S} \right) \tag{10.3}$$

and γ_1 is the noise factor.

© The Author(s) 2016
J.-S. Yuan, *CMOS RF Circuit Design for Reliability and Variability*,
SpringerBriefs in Reliability, DOI 10.1007/978-981-10-0884-9_10

Fig. 10.1 Schematic of a double-balanced Gilbert mixer

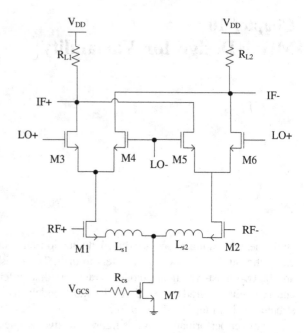

The sensitivity of the Gilbert cell mixer can be examined. The process variation and the aging effect may degrade the mixer performance. The conversion gain variation is modeled by the fluctuation of g_m and bias current drift as

$$\Delta CG = \frac{\partial CG}{\partial g_m} \Delta g_m = \frac{\partial CG}{\partial g_m} \left(\frac{\partial g_m}{\partial V_T} \frac{\partial V_T}{\partial I_{\text{bias}}} + \frac{\partial g_m}{\partial \mu_n} \frac{\partial \mu_n}{\partial I_{\text{bias}}} \right) \Delta I_{\text{bias}} \tag{10.4}$$

Expanding the partial derivatives in (10.4) the conversion gain variation can be written as

$$\Delta CG = \frac{2}{\pi g_m^2} \frac{R_L}{\left(R_S + \frac{1}{g_m} \right)^2}$$
$$\left\{ \frac{I_{\text{bias}}}{(V_{GSM1} - V_T)^2} \frac{L}{\mu_n C_{ox} W_{CS} (V_{GSCS} - V_T)} + \frac{I_{\text{bias}}}{\mu_n (V_{GSM1} - V_T)} \frac{2L}{C_{ox} W_{CS} (V_{GSCS} - V_T)} \right\} \Delta I_{\text{bias}} \tag{10.5}$$

where V_{GSM1} is the gate-source voltage to the RF transistor and V_{GSCS} is the gate-source voltage to the current source transistor.

Fig. 10.2 Process insensitive current source

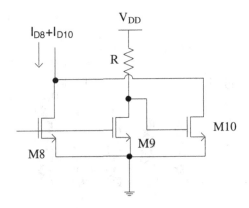

Similarly, the noise figure drift is derived as

$$\Delta F = \frac{\partial F}{\partial g_m}\Delta g_m = \frac{\partial F}{\partial g_m}\left(\frac{\partial g_m}{\partial V_T}\frac{\partial V_T}{\partial I_{bias}} + \frac{\partial g_m}{\partial \mu_n}\frac{\partial \mu_n}{\partial I_{bias}}\right)\Delta I_{bias}$$

$$= \left\{\frac{\pi^2}{4}\left(\frac{-2\gamma}{g_m^2 R_S} - \frac{1}{g_m^3 R_L R_S}\right)\right\}$$

$$\left\{\frac{I_{bias}}{(V_{GSM} - V_T)^2\,\mu_n C_{ox} W_{CS}(V_{GSCS} - V_T)} + \frac{I_{bias}}{\mu_n(V_{GSM} - V_T)}\frac{2L}{C_{ox} W_{CS}(V_{GSCS} - V_T)}\right\}\Delta I_{bias}$$

$$(10.6)$$

Equations (10.5) and (10.6) account for process variations and aging effect of the mixer.

It is clear from (10.4) to (10.6) that the mixer performance is dependent on the drain current of current source. To maintain the mixer performance, the drain current of $M7$ has to be kept stable. Thus, the process invariant current source circuit shown in Fig. 10.2 is employed. In Fig. 10.2, drain currents of $M8$ and $M9$ are designed the same. Changes in $M8$ and $M10$ drain currents are negatively correlated to remain as a stable bias current ($I_{D8} + I_{D10}$). For example, if the process variation increases the threshold voltage, which decreases the drain current of $M8$, the gate voltage of $M10$ increases ($V_{G10} = V_{DD} - I_{D9}R$). Thus, the drain current of $M10$ increases to compensate the loss of I_{D8}.

ADS simulation is used to compare the mixer performance using the single transistor current source versus process invariant current source [2]. The RF mixer is operated at 900 MHz with an intermediate frequency of 200 MHz. In the circuit design, CMOS 0.18 μm mixed-signal technology node is used. R_{L1} is 210 Ω and R_{L2} is 190 Ω. The transistor channel width of $M3$–$M6$ is 200 μm. The channel widths of $M1$ and $M2$ are 190 and 210 μm, respectively. L_{s1} and L_{s2} are chosen at 2 nH. The width of $M7$ is 250 μm. The gate resistor size of the current source is 400 Ω. The mixer sets the gate biasing voltage at the current source at 0.62 V. In the current source, the transistor $M8$ and $M9$ match each other as 100 μm. The width of $M10$ is 600 μm. The supply voltage V_{DD} is 1.8 V.

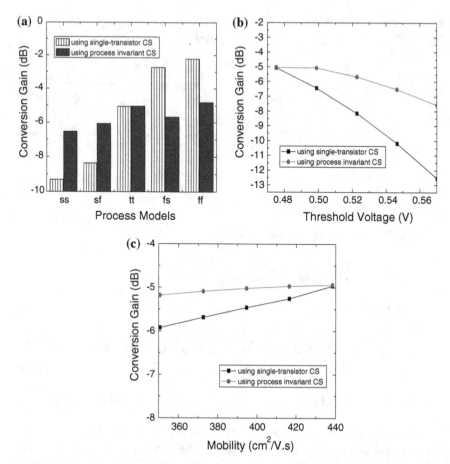

Fig. 10.3 **a** Conversion gain predicted by different process models. **b** Conversion gain versus threshold voltage. **c** Conversion gain versus electron mobility

For the process variation effect, the conversion gain of the mixer is evaluated using different process corner models due to inter-die variations. The simulation result of the fast-fast, slow-slow, slow-fast, fast-slow, and normal-normal models is shown in Fig. 10.3a. It is clear from Fig. 10.3a that the mixer with the invariant current source shows robust conversion gain against different process variations.

The conversion gain is also evaluated using different threshold voltage and mobility degradations resulting from aging (hot carrier effect) as shown in Fig. 10.3b, c. The hot carrier injection increases the threshold voltage, but decreases the electron mobility. The conversion gain decreases with an increased threshold voltage or decreased mobility due to reduced transconductance. Again, the mixer with process invariant current source exhibits more robust performance against threshold voltage increase and mobility degradation.

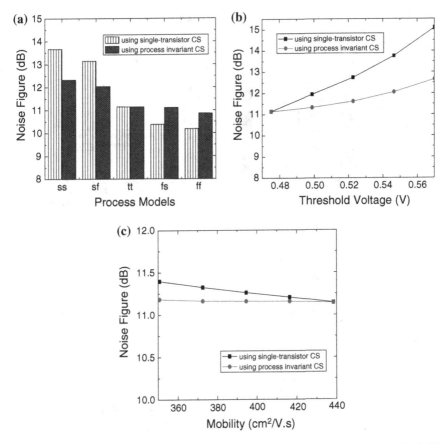

Fig. 10.4 **a** Noise figure predicted using different process models. **b** Noise figure versus threshold voltage. **c** Noise figure versus electron mobility

In addition, the noise figure of the mixer using the process invariant current source is compared with that using the single transistor current source. The noise figure versus different process models is displayed in Fig. 10.4a. It is clear from Fig. 10.4a that the noise figure is more stable over different corner models for the mixer using the current invariant current source. The noise figure also shows less threshold voltage and mobility sensitivity as evidenced in Fig. 10.4b, c. In Figs. 10.4b and 10.5c, the noise figure increases with increased threshold voltage and decreased mobility due to reduced drain current and transconductance in the mixer.

The output power of the mixer has been evaluated using different process corner models as well. As shown in Fig. 10.5a the output power of the mixer using the process invariant current source demonstrates robust performance against process variations. In Fig. 10.5b, c the output power decreases with increased threshold voltage and decreased mobility due to reduced drain current in the mixer. The

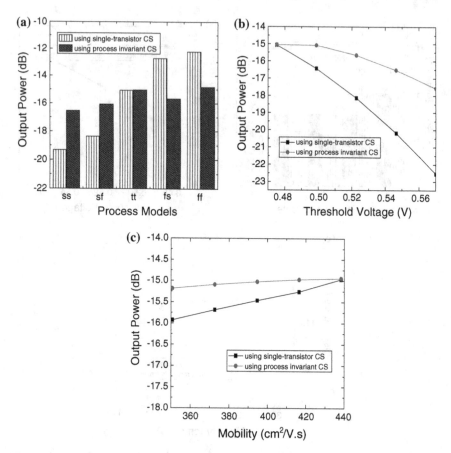

Fig. 10.5 **a** Predicted mixer out power using different process models. **b** Output power versus threshold voltage. **c** Output versus electron mobility

output power in Fig. 10.5b, c also shows less sensitivity against aging effect, which increases the threshold voltage and decreases the electron mobility.

The output power of the mixer has been evaluated using different process corner models as well. As shown in Fig. 10.5a the output power of the mixer using the process invariant current source demonstrates robust performance against process variations. In Fig. 10.5b, c the output power decreases with increased threshold voltage and decreased mobility due to reduced drain current in the mixer. The output power in Fig. 10.5b, c also shows less sensitivity against aging effect which increases the threshold voltage and decreases the electron mobility.

To further examine the process variation and reliability impact on RF mixer, Monte Carlo (MC) circuit simulation has been performed. In ADS, the Monte Carlo simulation [3] assumes statistical variations (Gaussian distribution) of transistor model parameters such as the threshold voltage, mobility, and oxide thickness. In the Monte Carlo simulation, a sample size of 1000 runs is adopted. Figure 10.6a, b

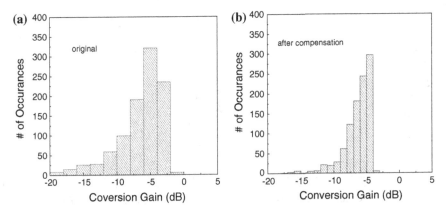

Fig. 10.6 a Conversion gain statistical distribution without compensation. **b** Conversion gain statistical distribution after process compensation effect

display the histograms of conversion gain using single transistor current source (original) and using the process invariant current source (after compensation). For the mixer using the traditional current source, the mean value of conversion gain is −6.608 dB and its standard deviation is 3.18 %. When the process invariant current source is used, the mean value of conversion gain changes to −6.324 dB and its standard deviation reduces to 2.08 %.

The noise figure after 1000 runs of Monte Carlo simulation is dialyzed in Fig. 10.7a, b. For the mixer using the single transistor current source, the mean value of noise figure is 11.667 dB and its standard deviation is 2.49 %. When the process invariant current source is adopted, the mean value of noise figure changes to 11.159 dB and its standard deviation reduces to 1.29 %. Clearly, the mixer using

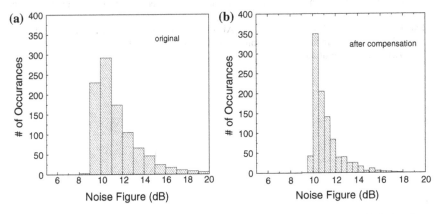

Fig. 10.7 a Noise figure statistical distribution without current compensation. **b** Noise figure statistical distribution after process compensation effect

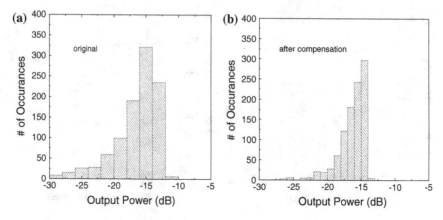

Fig. 10.8 a Output power statistical distribution without current compensation. **b** Output power statistical distribution after process compensation effect

the process invariant current source shows better stability against statistics process variations.

In addition, the output power of the mixer is examined in Monte Carlo simulation. Figure 10.8a, b demonstrates an improvement of output power for the mixer using the process invariant current source over that using the traditional current source. In Fig. 10.8, the mean value of output power changes from −16.608 to −16.324 dB and its standard derivation reduces from 3.81 to 2.08 % once the process invariant current source is used.

References

1. Lee TH (1998) The design of CMOS radio-frequency integrated circuits. Cambridge University Press
2. Pappu AM, Zhang X, Harrison AV, Apsel AB (2007) Process invariant current source design: methodology and examples. IEEE J Solid-State Circuits 2293–2302
3. http://www.agilent.com/find/eesof-ads

Printed in the United States
By Bookmasters